Global Economic Crisis and Local Economic Development

This book offers a collaborative investigation of the policies and practices which have redeveloped local and national economies in the aftermath of the global economic crisis which erupted in 2008. It explores 'localised' models of economic development, including problems of diversity and balance and the role of firms, industries and clusters, alongside comparative studies of policy responses to the crisis at local, regional and national levels.

Global Economic Crisis and Local Economic Development seeks routes for economic development in a post-crisis world. The roles of innovation, entrepreneurship, knowledge infrastructures, public policies, business strategies and responses, as well as global contexts and positioning, are explored as investigative themes which run throughout the collection as a whole. This text brings together a range of international disciplinary experts from economics, geography, history, business and management, politics and sociology. Its coverage is comparative and global, with contributions focusing on the USA, Japan, China and India, as well as European contexts and cases.

This book is of value both for the intrinsic quality of its individual studies and for the contrasts and comparisons enabled by the collection when viewed as a whole. It has an accessible but rigorous style, making it ideal for a range of users including academics, researchers and students who study economic development and regional development.

Jason Begley is Research Fellow at Centre for Business in Society and is Head of the Motor Industry Observatory (MIO), Coventry University, UK.

Dan Coffey is Graduate School Director and Senior Lecturer at Leeds University Business School, University of Leeds, UK.

Tom Donnelly is Professor of Automotive Business at Coventry University Business School, Coventry University, UK.

Carole Thornley is Emeritus Professor of Employment and Public Policy at University of Keele, UK.

Global Economic Crisis and Local Economic Development

International cases and policy responses

**Edited by
Jason Begley, Dan Coffey, Tom Donnelly
and Carole Thornley**

 Routledge
Taylor & Francis Group

LONDON AND NEW YORK

First published 2016
by Routledge
2 Park Square, Milton Park, Abingdon, Oxon OX14 4RN

and by Routledge
711 Third Avenue, New York, NY 10017

Routledge is an imprint of the Taylor & Francis Group, an informa business

British Library Cataloguing in Publication Data
A catalogue record for this book is available from the British Library

Library of Congress Cataloging in Publication Data
Names: Begley, Jason, editor. | Coffey, Dan, 1966- editor.
Title: Global economic crisis and local economic development : international cases and policy responses / edited by Jason Begley, Dan Coffey, Tom Donnelly and Carole Thornley.
Description: New York : Routledge, 2016.
Identifiers: LCCN 2015040928 | ISBN 9780415870368 (hardback) | ISBN 9781315630328 (ebook)
Subjects: LCSH: Global Financial Crisis, 2008–2009. | Economic development. | Economic policy. | Organizational change.
Classification: LCC HB3717 2008 .G5628 2016 | DDC 338.9—dc23LC record available at http://lccn.loc.gov/2015040928

ISBN: 978-0-415-87036-8 (hbk)
ISBN: 978-1-315-63032-8 (ebk)

Typeset in Times New Roman
by FiSH Books Ltd, Enfield

MIX
Paper from responsible sources
FSC
www.fsc.org FSC® C013604

Printed and bound by CPI Group (UK) Ltd, Croydon, CR0 4YY

Contents

Figures

Tables

Contributors

Jason Begley is Research Fellow at Centre for Business in Society and is Head of the Motor Industry Observatory (MIO), Coventry University, UK.

Michael H. Best is Emeritus Professor of Economics at the University of Massachusetts Lowell, USA and Babbage Fellow, Babbage Industrial Policy Network Institute for Manufacturing, Cambridge University, UK.

J. Robert Branston is Associate Professor in Business Economics and Deputy Director of the Centre of Governance and Regulation at the University of Bath School of Management, UK.

Holger Bungsche is Professor of Economics at the School for International Studies, Kwansei Gakuin University, Japan.

Dan Coffey is Graduate School Director and Senior Lecturer at Leeds University Business School, University of Leeds, UK.

Keith G. Cowling is Emeritus Professor of Economics at the University of Warwick, UK.

Anthony P. D'Costa is a Professor in the School of Social and Political Sciences at the University of Melbourne, Australia.

Tom Donnelly is Professor of Automotive Business at Coventry University Business School, Coventry University, UK.

Helen Lawton Smith is Professor of Entrepreneurship in the Department of Management at Birkbeck, University of London, UK.

Jonathan Michie is Professor of Innovation and Knowledge Exchange at the University of Oxford where he is Director of the Department for Continuing Education and President of Kellogg College.

Minh-Trang Thi Nguyen is an Academic Tutor at Wake Forest University, USA.

Christine Oughton is Professor of Management Economics and Head of the Department of Financial and Management Studies at the School of Oriental and African Studies (SOAS), University of London, UK.

James M. Rubenstein is Professor of Geography at Miami University, USA.

Ian M. Taplin is Professor of Sociology and International Studies at Wake Forest University, USA.

Carole Thornley is Emeritus Professor of Employment and Public Policy at University of Keele, UK.

Philip R. Tomlinson is Associate Professor in Business Economics at the University of Bath School of Management, UK and a Convenor for the Institute for Policy Research (IPR).

Rupert Waters is Visiting Research Fellow at the Department of Management, Birkbeck, University of London, UK, a PhD student at Halmstad University's Department of Marketing, Sweden, and a local government officer in Buckinghamshire, UK.

James R. Wilson is Senior Researcher at Orkestra-Basque Institute for Competitiveness and Lecturer at Deusto Business School, University of Deusto, Spain.

1   Introduction

Jason Begley, Dan Coffey, Tom Donnelly
and Carole Thornley

This collection of studies is a collaborative investigation into the policies and practices that have sustained local communities in the context of the global economic threat which exploded in the economic and financial crisis of 2008–2009 (the first manifestations of which, particularly in the USA, were already becoming evident in 2007), and which offer routes towards building (or rebuilding) secure models of economic development. For this purpose, it brings together a range of international experts drawn from a number of disciplinary backgrounds: economics, geography, history, business and management, politics and sociology. Case analyses and policy appraisals carry the reader through the issues facing local economic development in the wake of the crisis, married with empirical and conceptual overviews of policy strategies, including strategies for innovation and less fragile forms of economic development. The coverage is global, with contributions focusing on China, India, the USA and Japan, as well as European contexts and cases, with levels of analysis ranging from cities to the national and international.

Faced with the difficult realities, the guiding spirit of this collection is to accept that there is 'no single best way' which all regions or localities should follow. It is intended rather to add to the research and policy-oriented materials now being debated around the world as to questions of resilience and recovery in the wake of the crisis. In keeping with this, no attempt has been made by the editors to coax a 'consensus' of inferences or conclusions from its contributors, since the aim is not to fix upon a formula. A variety of perspectives are offered: conceptually and empirically grounded approaches to a complex terrain where policy actions have ranged from state-sanctioned programmes of rationalisation and restructuring, nationalisation and pump-priming, to policies for sustainable growth and innovation at the more local, regional or sectoral level. The balance of the book emphasises the latter, but alongside broader purviews.

The economic and financial crisis and its fallouts

Precisely when writers tend to date the crisis partly reflects the point at which it manifested most obviously in their own economies. But its general contours, since rehearsed many times, are now familiar ones. There was a preceding interval which even in its own time was being called 'the great moderation', a phrase that was popularised internationally by Ben Bernanke,[1] twice chair of the US Federal Reserve, to capture an empirical phenomenon of reduced year-on-year variability in national output indicators growing around a positive trend. This ostensibly began in the middle part of the 1980s. Events have since called for a rethink; but even in the midst of this interval we should recall not only that there were still sharp recessions in individual economies but that these years saw the South East Asian crisis, the travails of Japan, and so forth. It is not unjust to criticise proponents of the great moderation thesis not only for insensitivity to underlying problems, but also for prioritising America and Western Europe.

Be this as it may, the great moderation was posed partly as a question that hinged on the relevant sense of prosperous times. Explanations often ran in terms of structural changes successfully realised and improvements to institutions successfully accomplished, as well as new technologies. Other factors, owing more to serendipity than design, include an overlapping slump in oil and energy prices, the run-up to the crash being marked amongst other things by a surge in the trading value of oil. The prospects of new markets and access to resources, vis-à-vis the new policies of openness being pursued in India, China and parts of Latin America, and the fall of the Soviet Union, were also material factors, as too were developmental trajectories in North and South East Asia.

But nonetheless, by the time the crisis struck it was apparent that the Anglo-American model of liberalised capitalism had been riding a wave of rising household indebtedness, compensating an extended period of depressed wage growth to fuel a consumption binge for those able to access credit (sometimes on absurd and, in the case of American poor and minorities, discriminatory terms) on the back of housing bubbles. By the time the crisis entered its 'panic phase'[2] with the March 2008 bankruptcy of Lehman Brothers, and a series of major rescues including the likes of Merrill Lynch, American International Group (AIG), Freddie Mac and Fannie Mae in the USA, and while financial journalists struggled to explain debt swaps and derivatives to readers, there had already been major revelations that all was not well in American and European finance. As the scope of the crisis spread, it spilled over into real economy activity, with conditions best described as depressed growth following sharp contractions in the West, with differential impacts and national policy responses further away from the Western epicentres.

While national incomes have restored somewhat in the USA, Britain and Northern Europe, the southern expanses of the Eurozone region remained depressed by austerity; and even amongst seemingly recovered nations, best-case scenarios are notably fragile. Moreover, as the first shock of the crisis years of 2008–2009 has faded, and as familiarity has resigned commentary to Europe's unresolved problems, pessimism is already turning to the potential third wave in the series of shocks: the initial breakout of the financial crisis, and ensuing slump, being first; the crisis of the Eurozone second. Capital flight from emerging market economies, in a context of (once again) falling oil and commodity prices, and the eventual expected tightening of financial conditions – although the interest rates set by central banks remain, like the yield on bonds newly issued to fund government borrowing programmes, suppressed in both America and Europe – is seeing pessimistic notes struck by the World Bank. As its chief economist notes, in a survey pointing to difficulties for low- and middle-income countries: 'Global growth has yet again disappointed, especially but not surprisingly in oil exporters and large developing countries' (World Bank 2015: xiii). How things might look even a year after this present collection appears in print – whether better or worse – is difficult to guess.

Resilience and recovery

But even radical uncertainty of this kind is not an excuse for failure to reflect on questions of local economic resilience and recovery, and on national cases and contexts. As will quickly become apparent, the contributions in this collection are concerned not so much with the details of the crisis to date, or global prophecy on crises to come, as with what has provided or will provide resilience, promote recovery and promise sustainability on the basis of selections of evidence and cases to date. In keeping with this, the micro-studies aimed at local, regional or sectoral levels in this collection show accordingly a preoccupation with policy and with what makes policy effective – most particularly, but not exclusively, from the viewpoint of business growth. The other side of this, of course, is that the absence of the requisite combinations of factors that contribute to desired outcomes means that some localities or regions, whatever the initial impact of the crisis, have since struggled to recover more than others. These are complemented by broader overviews of policies and outcomes on the national or supra-national plane, but likewise focused on question of aptness and sustainability.

Some prominent threads run throughout the individual studies. There is, for example, an awareness of the abiding importance of manufacture, whether in business innovation, an area in which net investment is known to have potentially significant spill-overs, or in the case studies of particularly (or potentially) successful post-crisis localities.[3] The changing contours of the world economy and the balance of industrial strengths provide a backdrop, not least in the prominence given in a number of studies to Asia. This does not mean that services are neglected: diversity and balance are also recurring themes. The ways in which policy interventions affect growth – the role of firms, industries and clusters; entrepreneurship; knowledge infrastructures; skills; access to institutional support and resources – figure as prominently as one might expect. There are concerns too with inequalities, whether between regions or social groups. Another form of inequality highlighted is how differential access to central decision makers affects the ability of localities to leverage funding and make connections. The importance of further education, of graduates, of universities, is similarly evident, and likewise the central importance of the geographical dimension for policy issues.

While we will not attempt an exhaustive list, it is notable that the gendered aspects of policies and outcomes figure prominently in several of the studies, in each case in relation to the career prospects and rewards for women, explored in both instances in connection with a discussion of the deteriorating position of wage earners. The overall effect of economic crises and policy responses on the position of women is a critically important question, albeit one that is sometimes still easily overlooked.

Taken together, the studies in this collection offer a series of perspectives on some of the most important issues that are essential to effective public policy formulation in the aftermath of the shocks to hit the global economy. The question as to whether local or regional strategies offer meaningful solutions in a context of global travails will be admitted by all readers, whatever their own vantage points, to be a key one: do models of more local forms of economic development, as too models drawing on industry specific experiences, exist as ways forward in light of still ongoing economic trauma? There are also questions as to the appropriate levels upon which to understand local economic development in its practical contexts, and here a broader sense of the major issues and dilemmas facing economies on the bigger planes is also essential. Sensible policy must accept practical complexity, with multiple layers of causation. With this as a guiding motto for the studies which follow, we turn to the individual contributions.

Chapter studies and progression

The individual chapter contributions broadly divide into several categories in terms of their thematic positioning within the organisation of this book. The first set relate to questions of renewed or continued growth and innovation, whether applied to businesses at large or to specific localities, regions or industries. The second relate to comparative national and supra-national responses and contexts vis-à-vis the crisis.

Chapter 2 by Jonathan Michie and Christine Oughton ('Creating local economic resilience: co-operation, innovation and firms' absorptive capacity') explores what is required to promote innovation that will rebalance an economy, foster corporate diversity and create regional resilience. It points to the increased significance of 'resilient regions'. Like economic balance and economic diversity, resilient regions are to be seen as national assets in the context of global competitive pressures. Drawing on the lessons of their previous research, Michie and Oughton highlight the importance of cooperation and collaboration between firms for innovation – expanding expertise and developing specialist products in pursuit of corporate objectives – and of developing public policies conducive to this. This includes public policy to also promote cooperation between firms and universities, to facilitate technology transfers for products and processes. The research question posed for purposes of econometric investigation is 'whether innovation performance can be enhanced by co-operation and, if so, what determines firms' propensities to cooperate with universities'. The authors propose that in practice cooperation and innovation are jointly determined subject to the absorptive capacities of firms; absorptive capacity in turn is measured by firms' ability to obtain a positive pay-off from cooperation. Hence, in addition to promoting firm-university links, public policy must look to increase firms' absorptive capacity. This theme is rigorously explored using European Community Innovation Survey (CIS3) data.

In keeping with the observation that an assessment of what makes for success may point too towards factors that help explain failure, the authors of this opening study identify what they describe as a 'regional innovation paradox'. This is that lagging regions, where firms have the greatest need for innovation, are also regions whose firms are likely to demonstrate the lowest capacity to absorb the potential benefits of policy support, engage in collaborative innovation activities and exploit relationship incentives.

Chapter 3 by Rupert Waters and Helen Lawton Smith ('Global economic crises and local fortunes: the case of Oxfordshire and Buckinghamshire) explores how two dissimilar local economies in England have fared since the crash. Despite challenging times, each of these localities – Buckinghamshire and Oxfordshire – has in fact done well, despite some marked differences explored by the authors: in business demography, public sector employment and research assets. Drawing on comparisons based on national data-sets, the authors ask what explains their similarly strong performance notwithstanding such differences: Why has each proved resilient, notwithstanding obvious disparities in the resources open to them? In this respect Waters and Lawton Smith contrast positive policy responses and policy evolution, assessing the comparative exploitation in each locality of government-sponsored but locally rooted economic partnerships bringing together businesses, local government and other interested parties. It contrasts the strategies adopted by Oxfordshire, with its large and internationally renowned education base and substantial public sector (including the University of Oxford), as well as significant private sector employers in areas including automotive, with that of Buckinghamshire, less initially successful in winning funding from official programmes and, lacking Oxfordshire's advantages, reliant to a much greater extent on micro-businesses, but contriving nonetheless to create 'a unique set of economic development actors'.

Waters and Lawton Smith emphasise that both of these English regions, despite manifest differences, have not only done historically well in absolute terms, but are now pushing ahead of other British regions that have suffered more from the crisis. As with the opening study in this collection, knowledge-intensive sectors and knowledge infrastructures are also considered important, including for manufacture, as too levels of connectedness shown between firms and other key institutional actors. Another main lesson drawn is the significance of inherited industrial structures – of clustering and agglomeration, with associated patterns of entrepreneurship and innovation as well as access for businesses to resources and links with institutions.

Chapter 4 by Michael Best ('Regional capabilities and industrial resiliency: specialization and diversification dynamics in Lowell, Massachusetts') is a fascinating and highly detailed chapter which further explores similar questions of regional capabilities and industrial resilience, in an original and substantive fashion. Employing the case study of specialisation and diversification dynamics of regionally co-located high-tech companies in Lowell, Massachusetts, this uses a 'dynamic capabilities perspective'. Here, Lowell's high-tech companies serve as an industrial laboratory to examine business specialisation and diversity, aided by an historical database of high-tech companies designed to discover regional production and technology capabilities. For Best, the experiences of Massachusetts in resilience and renewal continue to confound many, and require an understanding that the regional growth process is more complex and interactive than is commonly acknowledged. He demonstrates that today's high-tech companies are members of specialised groups that both leverage and renew long-established regionally distinctive production and technology capabilities and skills. More specifically, the region in question is particularly strong, today as yesterday, in instrument making, machine tools and industrial equipment. Renewal and self-organising scale-up processes are illustrated with recently established groups of network communication equipment and software tools companies.

It is proposed that both groups illustrate the contribution of an 'open systems' business model to the renewal of regional production and industrial innovation capabilities which underlie the Massachusetts region's resilience to the pressures of global competition. One notable feature in this highly thought-provoking chapter, which engages with subtle literatures in subtle ways, and which will also undoubtedly generate wider interest in the impact of local histories and cultures on local economic development, is the argument that economic downturn can actually be functional to regional renewal processes – an imaginative and vital insight in the context of global economic dislocations.

Chapter 5 by Ian Taplin and Minh-Trang Thi Nguyen ('From recession to re-industrialisation: a case study of employment changes in North Carolina') similarly explores the renewal of domestic US manufacture, while confronting a worrying dilemma that faces many of America's Southern states: Is this overly dependent upon low entry-level hourly wages, low tax rates, union-free sites, and state and local government subsidies, or do other avenues for recovery present themselves? To this end, the authors explore structural change in North Carolina, a Southern state historically prone to the pursuit of low-wage business strategies, and with particular regard to the revival of its manufacturing fortunes in its traditional industries of textiles and furniture, and in a range of other sub-sectors including household goods and appliances and precision instruments and equipment. In the broader context of America's prospects for reindustrialisation, they criticise wage-depressing strategies as sitting badly with the needs for sufficiently qualified workers possessing at least basic engineering and computer literacies and post-secondary levels of education. The critical issue, as Taplin and Nguyen see it, is whether the incentives packages offered by North Carolina to attract inward state investment will combine with the emergence of dominant high-road strategies – in which businesses provide more extensive worker training, in partnership as needs be with local community colleges and other institutional actors and incentivised through higher employee wages, autonomy and workplace involvement to underpin retention of skills – or whether the costs and the risks will prove too unattractive, and wage depression will continue.

While manufacture is again pinpointed as a key sector, full play too is given to the spread of economic activity across industries: finance and insurance, health care, real estate, rental and leasing, and agriculture are also essential to the state economy. But it is intriguing to note the weight that Taplin and Nguyen give to players on the international scene in co-determining the manufacturing prospects of the American state: the impact of the Asian export-led model of economic growth on North Carolina – including Chinese and Indian textile firms – runs as a thread through the study as a whole. In competition with America's rust-belt, and taking advantage of proximity to America's eastern seaboard and southern sun-belt markets, North Carolina has been directly impacted by the evolving balance of international advantages, another policy concern. This study is also notable for the weight given to changes in female employment prospects.

Chapter 6 by Anthony D'Costa ('Institutional change, industrial logics, and internationalization: growth of the auto and IT sectors in India') focuses on two sector-level studies for the Indian subcontinent: information technology (IT) and automotive. This takes us to the world as it seems from the other side, but looking not so much at the traditional staples of the Indian economy like textiles, as at newer successes. The case study sectors are chosen not only for their success but for their differences: one (focusing mainly on software services) is a highly tradeable service, the other (focusing on passenger cars) a high-value-added consumer durable manufacture. While developments in each represent seeming departures from earlier industrial trajectories in India, D'Costa explains how successes in each are only partly the outcome of institutional change in the form of new rules governing how the Indian economy works, and its increased integration into the international trading economy: there are also particular initial conditions and industrial logics at play that have also been essential to the dynamics of growth in each of these two sectors. D'Costa favours a vantage point that is influenced by a longer evolutionary perspective, and which acknowledges the Indian state as a critical and essential actor, initially protecting India's industrial sectors from exposure to the world market but gradually 'reconstituting itself'. In the course of this careful investigation, the author leads the reader though contexts, analysis of markets, industry structures and business models, foreign direct investment, outsourcing and the positive linkages between industrial development and macro-economic growth. The study also draws attention to and explores the impact of India's diaspora on its prospects.

If the Indian case studies serve as further conceptual and empirical contributions to the themes explored in the opening set of chapter contributions, they also bridge a change in focus in the second half of this book, towards questions of how national economies have positioned their overall responses to problems of economic dislocation. In D'Costa's study, the 'big picture' emerges in the contexts informing the complex trajectories of two rising industrial sectors that have helped change the face that India presents to the world. In the studies to follow, the exploratory lens is more squarely directed towards explorations of broader comparative features: for Japan, China, America and Europe.

Chapter 7 by Holger Bungsche ('A system abandoned: 20 years of management, corporate governance and labour market reforms in Japan) begins with business school interest in Japan from the 1980s, highlighting the unique features of the Japanese company and management system and the organisation of the Japanese labour market which seemed to observers around the world to have played a prominent role in explaining Japan's successes in the international economic arena. Bungsche then turns to how Japan has responded to its own major economic crisis – one that preceded both the recent economic and financial crisis with its epicentre in the West, as indeed the earlier traumas of the South East Asian financial crisis, in 1997–1998. Here it needs only to be recalled that the Japanese bubble burst in the first part of the 1990s, which slowed growth some years before the regional crisis, during which Japan sharply recessed. The subsequent years have seen intermittent national recovery, punctuated by set-backs. The focal point of Bungsche's detailed and highly informative investigation, however, is how Japan has moved away from its own model for success, as a consequence of post-bubble reforms. A new framework, driven by pressures from 'business associations, companies, financial markets and other interest groups', represents a shift away from the company and management system, and labour market organisation, which made Japan famous, thus abandoning the old seniority system and main bank system which facilitated a national focus on long-term company growth and development. Bungsche describes a state of affairs in which a system has withered without an as yet coherent whole to replace it. The paths by which this has occurred are explained and the outcomes forensically explored, and not least for wages, non-standard employment forms and women workers.

Chapter 8, by Jason Begley and Tom Donnelly ('China and the global recession') turns to China, now contending with America to be the world's largest economy. They explore China's reaction to the global economic crisis, a response reflecting its unique economic, political and social structures. While the Chinese financial sector per se was (like Japan's) not hugely exposed to the immediate panic that beset American and European financial institutions, its real economy suffered from the loss of exports to the West and other affected trading partners. Running through a formidable array of policy tools and actions, they underline the vigour of the initial Chinese response, both monetary – with policy tools including quotas on domestic bank lending, central bank deposit requirements, restrictions on mortgage lending rates and down payments – and fiscal. It is interesting in this latter respect to consider the extent to which China adhered to a policy stance since largely abandoned in the richer Western economies, with direct fiscal stimulus aimed both at demand management and infrastructures. The abiding reach of the state is equally demonstrated in the breadth of actions extending to the pricing of household appliances purchased by low-income rural families and support not only for key industries but also (diminished) urban and rural medical services. At the same time, in a detailed overview, Begley and Donnelly highlight too the equally formidable array of challenges still facing China: economic imbalances with trading partners without, regional imbalances within; low rates of domestic consumption; an ageing population demographic;

severe problems of environmental degradation; and so forth. China is viewed as an economy in structural transition from a planned to a market economy, avoiding exposure to financial contagion but now sufficiently integrated with the global economy to be vulnerable to loss of export markets or inward foreign direct investment, albeit still able to command a plethora of policy tools by way of response.

It is interesting to compare some aspects of China's current challenges with those described also for the richer and more developed Japan. In Japan, the loss of one system has accompanied still unresolved social and economic trauma, justified in political and business circles as essential but with fruits yet to be demonstrated. In China, fear of social unrest, as a consequence not only of the shock of the Western crisis on rates of economic growth but also of severe domestic income inequalities and an underdeveloped social and welfare safety net, is propelling efforts to shore up the social infrastructures within a new market economy that in Japan have been unpicked within an old one. In each case, there are problems of economic insecurity and missing welfare nets. In each case, the national economy is threatened by an ageing demographic. In both instances, it was principally the real economy, including exports, which were hit by the Western economic and financial crisis, rather than the financial sector as such. But the changing contours of international economic relations in the Asia Pacific region also loom differently for each, with the rise of China as giant economic power a challenge to Japan as well as a business opportunity and source of markets and resources.

Chapter 9 by James Rubenstein ('North American and European responses to crisis in the motor vehicle industry') gives an excellent sense of the difficulties of political decision making for rapid responses in the midst of a major economic crisis. Along with a careful survey of the first impact of the crisis of 2008–2009, and the evolving situation subsequently, on the automotive industries of North America and Western Europe, it looks at contrasting policy responses. The result is another fascinating chapter. As loans to support motor vehicle purchase and production dried up, and credit lines to firms shrank, auto industry executives in North America made different assessments to their counterparts in Europe: in the former case, a view quickly gained ground that the industry was in structural crisis; in the latter, that there was a cyclical problem – the almost inevitable end of an unexpected long period of stability and now a return to previously familiar cycles in demand and production. As first efforts in America to consider mergers, quick fire asset sales, borrowing and state assistance floundered, policy moved to a combination of industrial and regional policy, with tacit outcomes not just for the USA but also for Canada and (looking further ahead) Mexico. Much of the weight of policy came down in the form of rationalisation and closure of facilities. By contrast, an initial general European recovery plan aimed (like China) at supporting demand, supplemented for automotive by various car 'scrappage' schemes, initially bolstered expectations of a lasting recovery that were dashed by subsequent experience. The chapter closes by considering not only the global contrast between responses on each side of the Atlantic, but at prospects for

Mexico, and controversy in the European Union (EU) over how much of the ongoing difficulties of the industry – albeit with national variations – is after all structural and how much due to car dumping from sites operating in Eastern Europe in a context where local markets have been depressed as a consequence of European austerity.

In all three of these successive chapters, domestic power structures and politics loom large: the support of particular social strata for the reforms which have done so much to terminate Japan's stakeholder capitalism; the support of China's communist party for its own hybridised capitalism; and the comparative difficulties and ease of major industrial concerns in realising political support in North America and Western Europe. And political as well as policy themes also emerge in the course of the study which follows.

Chapter 10 by J.R. Branston, Keith Cowling, Philip Tomlinson and J.R. Wilson ('Addressing "strategic failure": widening the public interest in the UK financial and energy sectors') closes the collection with a topical and timely chapter. It employs a strategic choice perspective to analyse what the authors perceive as 'underlying governance failures' exposed by recent global economic crises. The authors consider two case studies currently attracting public attention in Britain (as internationally) – the financial and energy sectors – whilst noting that their arguments could easily be extended to other important strategic sectors, such as health, education and transport. For Branston *et al.* the concentration of economic power and strategic decision making processes within key sectors has led to the spectre of 'strategic failure' and much public discontent. They argue that both current and proposed regulations are ineffective in delivering these (as, indeed, other) public services as services in the public interest, and that therefore what is required is a governance-centric approach that makes inclusion and democratic choice the focal points of public policy. They outline possibilities for aiding future developments in these sectors, for the wider public interest. This engaging chapter employs a wealth of theoretical and empirical analysis to argue for a wider transformation of the economy towards a 'more sophisticated form of economic democracy that is more responsive to a broader range of public interests and less likely to be captured by small elites of disproportionately powerful interests'. It is of considerable interest to policy makers and a wide range of readers looking for coherent and grounded approaches to local economic development in a context of economic uncertainties.

Notes

1   See Bernanke (2004).
2   We follow Elliot and Atkinson (2008: 40) and, more generally in these sentences, a survey of the crisis as it looked from Britain, from the viewpoint of immediate observation. The literature here is now massive, and international, and only partly overlapping in arguments.
3   One of the most interesting fall-outs of the crisis has been the move back to centre stage in policy discussions in both America and Western Europe of industrial policy: see, for example, for industrial policy in the EU, Gerlach *et al.* (2015) and for an inter-

national range of cases spanning both developed and developing economies, Stiglitz and Yifu (2013) and Stiglitz *et al.* (2013).

References

Bernanke, B.S. (2004) 'Remarks by Governor Ben S. Bernanke', Meetings of the Eastern Economic Association, Washington DC, February 20th 2004. See www.federalre-serve.gov/BOARDDOCS/speechES/2004/20040220/default.htm.

Elliot, L. and Atkinson, D. (2008) *The Gods That Failed: How Blind Faith in Markets Has Cost Us Our Future*, Random House: London.

Gerlach, F., Schietinger, M. and Ziegler, A. (eds) (2015) *A Strong Europe – But Only With a Strong Manufacturing Sector: Policy concepts and instruments in 10 EU member states*, Hans Böckler Stiftun: Schüren.

Stiglitz, J.E. and Yifu, J.L. (eds) (2013) *The Industrial Policy Revolution I: The Role of Government Beyond Ideology*, IEA Conference Volume, Palgrave Macmillan: Houndmills, Basingstoke and New York.

Stiglitz, J.E., Yifu J.L. and Patel, E. (eds) (2013) *The Industrial Policy Revolution II: Africa in the Twenty-First Century*, International Economic Association, Palgrave Macmillan: Houndmills, Basingstoke and New York.

World Bank (2015) *Global Economic Prospects: The Global Economy in Transition*, World Bank Group: Washington, DC. See www.worldbank.org/en/publication/global-economic-prospects.

2 Creating local economic resiliance

Co-operation, innovation and firms' absorptive capacity

Jonathan Michie and Christine Oughton

Introduction[1]

Following the 2008 financial crisis and the resulting global economic recession of 2009, there has been a renewed interest in 'rebalancing the economy', fostering corporate diversity (Michie and Oughton, 2013, 2014) and creating regional resilience (Christopherson *et al.*, 2010). All this is generally seen as being closely linked to and reliant upon fostering innovation, which has for some time been acknowledged to be critical for successful macroeconomic performance.[2] If anything, this belief in the importance of innovation has increased over the past few years since the global financial crisis, with governments striving to promote an innovative 'knowledge-based' economy. Likewise, it has been widely recognised for some time that collaboration is an important means of fostering innovation (see, for example, Dore, 1983; Teece, 1992; Smith *et al.*, 1995; Oughton and Whittam, 1997; Doz and Hamel, 1998; Ahuja, 2000; Love and Roper, 2001; Tether, 2002; Tether and Swann, 2003; Oxley and Sampson, 2004), but the regional aspects of this are attracting increasing attention as resilient regions become seen as important assets in the face of global competitive pressures. Such collaboration allows firms to expand their range of expertise, develop specialist products and achieve various other corporate objectives.[3] The importance of co-operation and networks of advice and information for successful innovation was detailed some time ago by, for example, the TRACES and HINDSIGHT projects in the USA and the SAPPHO project in the UK.[4]

The recognition that public policy to promote such co-operation is called for has, again, been increasing over the past few years. For example, firm-university co-operation was stressed by the UK Lambert Review (Lambert, 2003, commissioned by the government), which argued that university-firm collaborations facilitate technology transfer, which in turn encourages product and process innovation, and this enhances competitiveness and economic growth. The main policy conclusion of the review was that such collaboration should therefore be supported and encouraged. This chapter considers this prescription in the light of the evidence available from the European Community Innovation Survey (CIS3), using data for the UK.

One of the positive outcomes from such co-operation has been thought to be successful technology transfer.[5] As indicated below, the benefits, though, are not limited to technology transfer. Indeed, our results reported below would suggest that even to achieve successful technology transfer itself requires firms to undertake other strategic actions, some of which in turn are also likely to be assisted by co-operation with universities,[6] as well as with other bodies and firms.

The purpose of the current chapter is to contribute to the important theoretical, empirical and policy debates on this topic – of whether innovation performance can be enhanced by co-operation and, if so, what determines firms' propensity to co-operate with universities. Drawing on the literatures on innovation, co-operation and absorptive capacity, we find that co-operation and innovation are jointly determined and that firms need a certain degree of absorptive capacity in order to successfully collaborate over innovation activities. It follows that policies to promote innovation should incorporate both policies to promote firm-university links and policies to increase the absorptive capacity of firms.

Co-operation and innovation

The existing literature does demonstrate that innovation by firms is enhanced, or even depends upon, co-operation and collaboration, both between firms and other bodies such as universities, and networking between firms and their suppliers, customers or even competitors.[7]

There is evidence that forms of long-term relationship between independent firms may be superior to vertical integration as a means of co-ordinating the activities required for innovation, especially where these activities involve a high degree of technological 'strangeness' (Gomes-Casseres, 1994: 63). These new forms of alliance are prevalent in high-technology industries. There are indications that they contribute most to innovative performance when they involve a dense network of interpersonal relationships and internal infrastructures that enhance learning, unblock information flows and facilitate co-ordination by creating trust and by mitigating perceived differences of interest (Porter, 1990: 152–3; Moss Kanter, 1994: 97).[8]

These points regarding information flows were also brought out by Dore (1983) in his discussion of the 'obligated relational contracting' found between Japanese firms. This involves long-term trading relations in which goodwill (with 'give and take') is expected to temper the pursuit of self-interest, although this and other practices, notably labour market practices, have since come under strain, especially following the relatively slow economic growth of the 1980s and '90s.[9] In his 1983 article Dore argued that such relations were more common in Western economies than is generally recognised. While it may be objected that relational contracts lead to price-distortions and hence to a loss of allocative efficiency, they do lead to high levels of other kinds of efficiency. Specifically, 'the relative security of such relations encourages investment in supplying firms', 'the relationships of trust and mutual dependency make for a more rapid flow of information', and 'a by-product of the system is a general emphasis on quality'.

Kitson *et al.* (2003) used data from a number of surveys conducted by the Economic and Social Research Council (ESRC) Centre for Business Research (CBR), finding that, from the 1995 survey, half of the innovating firms had entered into collaborative partnerships, whereas only one in six of the non-innovating firms had entered into such arrangements. The overall impact of increased innovation and collaboration was found from the CBR data to be improvements in both output and employment growth rates – for individual businesses as well as for the economy as whole.[10] In terms of employment, fast-growth firms were almost twice as likely to have collaborated compared to firms with negative or no growth. Innovating firms were far less likely to have zero or negative employment growth than were non-innovating firms and were far more likely to have achieved fast growth in employment. The CBR survey data indicated a similar picture in the contrast between collaborators and non-collaborators – superior employment growth being shown by the collaborators. This superior performance of innovating firms and of collaborating firms was also apparent in terms of turnover growth and in terms of the growth of profit margins. Innovation and collaboration over the period 1987–91 were found to be statistically significant and positively correlated with innovating in the 1992–5 period.

The role of universities

Co-operation between firms and universities can enhance the innovative capacity and propensity of firms in a number of ways. First, such co-operation may allow firms to tap into the research skills and techniques acquired in university research.[11] Second, firms may gain access to what in some cases will be quite new bodies of scientific knowledge located originally, at least, almost entirely within universities.[12] And third, through the more general interaction which can assist in the development of new products and processes, solve problems and raise awareness of new possibilities. These aspects are all of course interlinked, and the relative balance of importance between them will alter not only between sectors and firms, and over time, but even over the lifetime of individual collaborative projects.

In the section below we start by examining how important, if at all, the evidence from the 3rd UK Community Innovation Survey (CIS3) suggests such co-operation to be, to the innovative performance of firms. Drawing on the concept of absorptive capacity (Cohen and Levinthal, 1990, 1994) we also focus on the determinants of firm-university co-operation. Of course this raises the possibility that innovation and co-operation are jointly determined, and so we test for this possibility using endogenous explanatory variables.

Evidence from the third UK Community Innovation Survey (CIS3)

CIS3 comprises responses from 8,172 business enterprises: it is a comprehensive survey of firms' innovation activities in the UK, providing data on various forms of direct innovation outputs and factors influencing innovation activities.[13] Prior to the CISs, most research into innovation had utilised patent data or data on R&D expenditures as a proxy for innovation output, and focused on high-technology-intensive manufacturing industries. The CIS provides direct measures of innovation output across all UK regions and most industry sectors. It also provides indicators of links between the university sector and industry and of knowledge transfer. The latest CIS survey also allows us to attain representative results on a range of innovation activities for UK regions.

Product innovation

Table 2.1 column (a) shows the total number of CIS3 respondents, column (b) the number of enterprises answering whether or not they had introduced any new or significantly improved products, and column (c) is the number of enterprises declared to have been product innovators. Column (d) gives the proportion of product innovators by UK region as a percentage of all enterprises answering the CIS question related to product innovation. Furthermore, all enterprises that were product innovators, listed in column (c), were asked if they engaged in *novel* product innovation, that is, introducing a new or improved product that was also new to the enterprise's market. The number of product innovators that were *novel* product innovators is reported in column (e) and the *proportion* of product innovators that this represents is given in column (f).

Knowledge transfer: co-operation agreements between university and business

The CIS allows us to gain a picture of the extent of co-operation and knowledge transfer between the university sector and industry. CIS3 allows us to identify co-operation between businesses, on the one hand, and universities and other higher education institutes and government research organisations, on the other. Table 2.2 reports the number and proportion of enterprises that had collaborative

Table 2.1 The number and proportion of *product innovators* and *novel product innovators*

	CIS 3 responses	... which answered the relevant question	... which were product innovators		... which were novel product innovators		
				$\frac{(c)}{(b)} 100$		$\frac{(e)}{(c)} 100$	$\frac{(e)}{(b)} 100$
	(a)	(b)	(c)	(d)	(e)	(f)	(g)
	No.	*No.*	*No.*	*Per cent*	*No.*	*Per cent*	*Per cent*
UK	**8,172**	**8,141**	**1,745**	**21.4**	**773**	**44.3**	**9.5**

Source: own calculations from UK CIS3 survey

Table 2.2 Enterprises with co-operation arrangements on innovation

CIS 3 responses	... which answered the relevant question	... which co-operated with universities		... which co-operated with government research organisations		... which co-operated with private research institutes	
		$\dfrac{(c)}{(b)}100$		$\dfrac{(e)}{(b)}100$		$\dfrac{(g)}{(b)}100$	
(a)	(b)	(c)	(d)	(e)	(f)	(g)	(h)
No.	*No.*	*No.*	*Per cent*	*No.*	*Per cent*	*No.*	*Per cent*
UK 8,172	**843**	**306**	**36.3**	**113**	**13.4**	**109**	**12.9**

Source: own calculations from UK CIS3 survey

innovation project(s) with universities, governmental institutes and private research organisations.

The first point to note is that the number of businesses that actively participate in joint innovation projects with other organisations is low.[14] Only 843 enterprises stated that they participated in a co-operative innovation agreement with other organisations, and only 306 reported that they have joint innovation projects with universities. Out of the co-operating firms, 36 per cent engaged in a partnership with universities or other higher education institutes and around 13 per cent collaborated with government organisations and private research institutes. The remainder co-operated with other firms or commercial laboratories/R&D enterprises.

The second point to note is that co-operation between firms and universities (and research organisations) appears to have a significant impact on the probability of a firm being an innovator. Of those firms that do not collaborate with universities, the probability of them introducing a new product is 19.5 per cent. For those firms that have joint innovation projects with universities, this probability increases to 71.6 per cent. This provides statistical support for the Lambert review's conclusions that business-university collaborations are associated with successful innovation and thus bring broader economic benefits.

Co-operation with universities and innovation

Table 2.3 reports the number of firms that had or did not have a co-operation agreement with a university and, in each case, the number of those that introduced either a product or process innovation.

Of those firms that had a co-operation agreement, 83 per cent innovated. Of those firms with no such co-operation agreement, only 28 per cent innovated. This association between co-operation and innovation is highly significant statistically.[15]

Table 2.3 Firms that had a university co-operation agreement and innovated

	Innovators	*Non innovators*	*Total*
University co-operation agreement	250	52	302
No university co-operation agreement	2157	5636	7793
Total	2407	5688	8095

Source: own calculations from UK CIS3 survey

What determines co-operation?

The above results show that co-operation between firms and universities increases the probability of a firm being an innovator. This conclusion leads to two further, inter-related questions. First, what determines such co-operation and what determines how successful it will prove? Second, given the clear evidence that co-operation pays, why is it not more widespread? These questions are related, of course, because one of the reasons why some firms are failing to pursue such arrangements may be because they see no clear benefit. And in some cases there may be a valid basis for such a view, either because the firm has pursued such policies itself in the past, or has witnessed others having done so, with no clear pay-back. Thus, the benefits of such arrangements may not flow automatically; their efficacy will depend on other actions taken by the parties in question. To the extent that firms fail to secure the full potential from such arrangements, there will be a concomitantly lower incentive to co-operate.[16]

In other words, the low numbers of collaborative agreements between businesses and universities and the differences across regions (with the South East having the highest number) may be due in part to differences in the *absorptive capacity* across firms (Cohen and Levinthal, 1990, 1994; Tether and Swann, 2003). Within the regional innovation literature this has been termed the regional innovation paradox (Oughton *et al.*, 2002).[17] The regional innovation paradox refers to the contradiction between the greater *need* for lagging firms or regions to innovate and their lower *capacity* to absorb public funds earmarked for the promotion of innovation or to engage in innovation related activities, including in co-operation with research partners such as universities. The regional innovation paradox (and the CIS3 data on the number and impact of joint innovation agreements between firms and universities across the regions) suggests a need for regional innovation policies to catalyse business-university links within the UK regions as a means of promoting innovation. There may be a similar paradox when it comes to individual firms, namely that those in most need of technology transfer from universities may have inadequate absorptive capacity to realise such transfers.

Mora-Valentin *et al.* (2004) analyse the impact of a series of contextual and organisational factors on the success of 800 co-operative agreements between Spanish firms and research organisations, finding factors such as commitment and trust to be important. The definition of objectives is also found to be important, which might appear contradictory to the trust factor, as one might think that these are alternative ways of conducting co-operation – either through mutual trust, or else through clearly defined objectives. Thus, detailed contracts might be thought of as an alternative way of avoiding opportunistic behaviour in the absence of trust.

Mora-Valentin *et al.*'s findings are actually consistent, though, with the findings of research from the ESRC's 1992–7 Contracts and Competition Programme. This reported that, where inter-firm relations were indeed based on a high degree of trust, agreements were nevertheless often specified clearly in written contracts, even though the reported assumption was that such contracts were to go in the drawer, with no expectation of having to refer to them. This finding, that trust on the one hand, and detailed agreements on the other, might be complements rather than substitutes, was nicely illustrated by one senior manager interviewed by one of the Contracts and Competition Programme's researchers, who reported that he always advised his new managers, 'never sign a contract with someone you don't trust'.[18]

Firms' absorptive capacity

We therefore looked at the CIS3 data to see whether the likelihood of firms co-operating with universities was connected to their absorptive capacity. The key five factors for creating or allowing absorptive capacity we took to be the size of the firm, the proportion of scientists and engineers (of degree level or above) in the workforce, whether advanced management techniques had been introduced, internal R&D expenditure and expenditure on training. The dependent variable was a binary variable indicating whether or not firms had a co-operative agreement for innovation with universities. The equation was estimated using a Probit regression model. Our results are reported in Table 2.4.

The results indicate that each of the five aspects of absorptive capacity has a significant positive association with the likelihood that a firm will have a co-operative agreement with a university.

The determinants of innovation

The literature on innovation suggests that there are a number of factors that explain differences in innovation performance across firms; these include the resource capability of the firm (Winter, 1987), its absorptive capacity (Cohen and Levinthal, 1990, 1994) and the extent of inter-firm and inter-organisational linkages (Oughton and Whittam, 1997). However, as the above discussion has indicated, co-operation and networking between firms is also partly determined by the firm's absorptive capacity – which includes its track record on innovation and its ability to innovate. This suggests that any analysis of the determinants of innovation should take into account the joint determination of innovation and co-operation.

To this end we specify a simultaneous relationship between innovation and co-operation, with co-operation as an endogenous explanatory variable. Our measure

Table 2.4 Determinants of firms' co-operation with universities: Probit regression

Dependent variable: co-operation with universities (binary dependent variable)

Explanatory variables	Coefficients	t-ratio	Marginal effects
Constant	−2.528**	−25.14	
Size of enterprise	0.176**	5.48	0.018
Science and Engineering graduates as a proportion of the workforce	1.289**	7.74	0.131
Expenditure on training	0.407**	5.11	0.044
Internal R&D	0.791**	10.03	0.112
Advanced management techniques	0.079*	1.94	0.008
N	2794		
Pseudo R^2	0.197		
Chi-squared	309.13		

Notes: ** denotes significance at 1% level
     * denotes 5% significance level

Source: own calculations

of innovation is the share of turnover derived from new products – this measure is preferred to binary measures of innovation (for example, whether a firm has introduced a new product or not) as it contains more information. We use a generalised Probit model with endogenous regressors, and our explanatory variables are:

i.   size of the firm
ii.   the proportion of scientists and engineers (of degree level or above) in the workforce
iii. expenditure on training
iv. expenditure on marketing
v.   expenditure on R&D
vi. expenditure on acquisition of machinery
vii. co-operation with universities (endogenous regressor).

Co-operation is instrumented and estimated as a function of received public support, advanced management techniques applied, enterprise size, proportion of scientists and engineers in the workforce, expenditures on training, marketing, internal R&D and machinery – that is, an equation very similar to that shown in Table 2.4. The endogenous equation is estimated as a linear function despite the fact that co-operation is a binary variable, since, as Angrist and Krueger (2001) state, it is not necessary to use Probit or Logit estimation to attain the predicted values from the endogenous equation and it 'may even do some harm … a linear regression for the first stage estimates generates consistent second-stage estimates even with a dummy endogenous variable' (Angrist and Krueger, 2001: 80). The results are reported in Table 2.5.

It can be seen that all of the explanatory variables take the expected sign and are significant at the 1-per-cent or 5-per-cent level with the exception of invest-ment expenditure on machinery which is significant at the 10-per-cent level.

We also tried including a variable that captured inter-firm co-operation among the explanatory variables, but this made identification of the separate effects of university-firm co-operation on the one hand, and inter-firm co-operation on the other, difficult as a result of these co-operation variables being strongly and significantly correlated. Hence, we estimated the same equation, this time leav-ing out the variable on university co-operation and replacing it with a variable measuring inter-firm co-operation. The results are presented in Table 2.6. Comparing the two equations, it can be seen that the results are similar, but a number of variables that are significant when firm-university co-operation is used as an endogenous regressor become insignificant when inter-firm co-operation is used. However, the effects of intra-mural R&D expenditure, the proportion of science and engineering graduates in the workforce and expenditure on market-ing new products all remain significant.

Table 2.5 Determinants of innovation and business-university co-operation: instrumental variables Probit regression with firm-university co-operation as an endogenous explanatory variable

Dependent variable: share of turnover from new products

Explanatory variables	Coefficients	z (t-ratio)	Marginal effects
Constant	−1.174***	−13.65	
Co-operation with universities	1.276***	2.63	0.475
Size of enterprise	0.087***	3.09	0.031
Science and Engineering graduates as a proportion of the workforce	0.414**	2.17	0.424
Expenditure on training	0.149***	2.48	0.054
Expenditure on marketing	0.662***	10.13	0.231
Internal R&D	0.375***	4.17	0.140
Expenditure on machinery	0.103*	1.62	0.037
N	2734		
Pseudo R^2	0.122		
Chi-squared	428.81		

Notes: *** denotes 1% significance level
** denotes 5% significance level
* denotes 10% significance level
Regression method is Probit with the endogenous regressor (co-operation with universities) estimated as a linear function

Source: own calculations

Table 2.6 Determinants of innovation and business-business co-operation: Instrumental variables Probit regression with inter-firm co-operation as an endogenous explanatory variable

Dependent variable: share of turnover from new products

Explanatory variables	Coefficients	Significance level	Marginal effects
Constant	−1.166**	−12.54	
Inter-firm co-operation	1.604*	2.38	0.575
Size of enterprise	0.046	1.12	0.017
Science and Engineering graduates as a proportion of the workforce	0.424*	2.11	0.153
Expenditure on training	0.090	1.19	0.032
Expenditure on marketing	0.575**	7.81	0.214
Internal R&D	0.322**	2.80	0.120
Expenditure on machinery	0.060	0.82	0.022
N	2723		
Pseudo R^2	0.122		
Chi-squared	427.89		

Notes: ** denotes 1% significance level
    * denotes 5% significance level
    Regression method is Probit with the endogenous regressor (co-operation with universities) estimated as a linear function

Source: own calculations

Discussion

The various phenomena reported on above – concerning co-operation and innovation – are clearly interrelated in various ways, with no doubt complex processes of simultaneous and joint causation. Co-operation and innovation are clearly correlated to a statistically significant degree. It does seem to be the case that co-operation can enhance innovation but, likewise, innovation may itself lead to co-operation; indeed, in some cases the decisions to co-operate and innovate will of necessity be simultaneous ones, where it is recognised that co-operation is a requirement for the intended innovation. It may be the decision to innovate that leads to co-operation, which in turn facilitates the innovation.

In addition to a complex causal relationship between these two phenomena, they may also be jointly influenced, caused or determined by other factors, such as corporate culture and the nature of the senior management in the firm. Some managers – for example those with university degrees, or even doctorates – may be more likely both to see co-operation as a natural method of the firm tapping into external knowledge and to see innovation as a necessary and inherent part of the firm's continued development and growth. In other words, those managers who tend to co-operate may also tend to innovate. In which case, the fact that the firm is both co-operating and innovating will be in part due to managerial culture.

This complex reality actually strengthens the key point being made above, namely that successful innovation requires more than just co-operation. Other factors are important, such as having an appropriately skilled, graduate work-force, conducting R&D and implementing advanced management techniques. The above results and discussion are also consistent with the literature on the management of innovation which stresses the need to increase corporate capacity for continuous change, and to approach innovation as not just technological development but also incorporating new work practices.[19]

The implications of this are first, that causal processes in this area will include two-way processes and joint causation. Policy as well as theory needs to recognise this complex reality. Second, there are likely to be virtuous circles that will be self-reinforcing once established, but which may require some critical mass of activity. Thus, Huggins (2001) explores the strengths and weaknesses of inter-firm network initiatives in the UK and finds a number of problems that have resulted in the level of such inter-firm interaction being fairly low and unintensive. This is despite the fact that these initiatives had resulted in substantial gains for the participating companies. The key was found to be sustainability of such networks.

Conclusion

The special issue of *Research Policy* on 'The Internationalisation of R&D' (see Niosi, 1999) was subtitled 'From technology transfer to the learning organisation'. As von Tunzelmann (2001) observes, there has been a shift of emphasis in modern accounts of technology transfer to, precisely, the conditions necessary for firms to absorb new knowledge on a continuing basis, as analysed in the growing literature on the 'learning economy', the 'learning region' and the 'learning organisation'.[20]

This chapter has reported evidence on the extent of university-firm collaboration and other co-operative arrangements between firms, on the impact that such collaboration and co-operation has on the propensity of firms to innovate, on the extent to which such innovation is new to the market and on the impact of such innovation in terms of the proportion of the firms' turnover accounted for by the resulting new and novel products. We find that the data do support the view that such co-operation is significantly correlated, to put it no more strongly, with the likelihood that firms will innovate.

In addition, the chapter has considered what the data can tell us about what other activities firms should be undertaking in order to enhance their innovative performance. Our results are consistent with much of the 'learning organisation' literature. We find that it is important not only for firms to co-operate with universities and others to access new knowledge, but also that firms need to have the necessary absorptive capacity if they are to benefit fully from the potential that such co-operation offers. To get the most from such arrangements, firms do need to be 'learning organisations'. This may require, amongst other things, a degree of in-house R&D, not only for the development of new products and processes by the firm, but also to have the capacity to benefit fully from other new knowledge that the firm is able to tap into through networking with others. It will require a skilled workforce that is able and willing to learn.

Conversely, any co-operation with universities needs to be with academics who are both able and motivated to achieve common objectives. Thus, for example, enhanced rewards for faculty involvement in technology transfer (based on a royalty distribution formula that determines the fraction of revenue from a licensing transaction that is allowed to go to a faculty member who develops the new technology) was found by Friedman and Silberman (2003) to be one of the factors that was positively correlated with university technology transfer. Siegel and van Pottelsberghe (2003) point out that this is broadly consistent with Link and Siegel (2002), which found that universities allocating a higher percentage of royalty payments to faculty members tend to be more efficient in technology transfer activities and that therefore it does appear 'that organizational incentives for university technology transfer are quite important' (Siegel and van Pottelsberghe, 2003: 6).

As far as firms are concerned, to gain full advantage from their relations with universities, firms need other things in place, such as a skilled, graduate workforce and an innovative corporate culture. How do firms ensure that these other requirements are met? In many cases the answer will be, in part at least, through successful networking with universities. In the case for example of requiring skilled personnel, this may be through tapping university brains on a continuing basis through consultancies and joint research projects. It may also be through the ability to hire the brightest graduates from the universities – graduates whom the firm may have been in contact with through the firm-university collaborative arrangements.

The importance of these types of activities in enhancing absorptive capacity and hence promoting innovation indicate why the local and regional economies remain vital even for firms operating globally, since many policies to build absorptive capacity have a regional dimension – for example, pools of skilled labour/graduates, training and firm-university interaction.[21] This is reinforced by the fact that local and regional co-operation tends to be associated with international co-operation rather than being a substitute for it.[22] Thus, while governments wish to promote innovation in order to enhance global competitiveness, many of the policy actions need to be delivered at a regional level. Indeed, the Lambert review called for significant new funding streams to be delivered through the Regional Development Agencies.[23] Such efforts to foster innovation regionally are supported by the Higher Education Innovation Fund, which is provided to universities in addition to funding for teaching and research.

Finally, to be successful, such policies need to overcome the innovation paradox referred to above. Unless this is done, then promoting university-firm co-operation will in many cases be like pushing a piece of string. The firms have to have the absorptive capacity to make use of such co-operation. This is not a reason for such firms to put off co-operating with universities; rather, it is drawing attention to the range of practices and policies that they need to be implementing, some of which will actually be facilitated by co-operation with universities. It is thus to point to the range of benefits that firms can enjoy from such co-operation. Being conscious of these will increase the likelihood that the benefits will actually be realised.

Appendix: the CIS3 dataset

The Office for National Statistics (ONS), with assistance from the Northern Ireland Department of Enterprise, Trade and Investment (DETI), conducted CIS3 on behalf of the Department of Trade and Industry (DTI) in 2001. The reference period is 1998 to 2000. The survey covers enterprises with ten or more employees in sections C-K of the UK Standard Industrial Classification of Economic Activities (UK SIC 92).[24] The unit of analysis is the enterprise.[25]

The sample was drawn from the Inter-Departmental Business Register (IDBR) divided into stratums by UK regions, size and industry sector.[26] In total *8,172 observations* were obtained which represents a 42 per cent response rate.[27]

Table 2A.1 gives the number of enterprises and the number of responses by UK region.

Looking at columns (c) and (g) of Table 2A.1, in which the number of responses and error margin are given, we find that regional returns for CIS3 are sufficiently high to allow statistically significant analysis at the 95 per cent confidence interval with an error margin of $\leq \pm 5$ per cent for all regions with the exception of Northern Ireland where the error margin is ± 5 per cent.

Columns (a)–(d) give the number and proportion of all enterprises and CIS responses by UK region. Columns (e) and (f) compare the number of valid CIS3 responses with the actual number of enterprises in each UK region. In terms of regions, the CIS sample is under-represented in the case of Northern Ireland, with a 35 per cent lower proportion of CIS responses compared to all enterprises. The North East, Scotland and Wales are over-represented.

We can determine the 95 per cent confidence interval for results from each region.[29] With 95 per cent confidence we are certain that the results in the CIS sample will be the same as in the UK population plus or minus the sampling error, also called the margin of error, reported in column (g) of Table 2A.1.

Notes

1   We are grateful for advice on econometric techniques from Professor Ron Smith and for help with the data analysis from Dr Marion Frenz.
2   In terms of the UK's economic performance, prospects and policy needs, see for example the various contributors to Yates (1992). In terms of the role of innovation and technology in economic growth more generally, see for example Mowery and Rosenberg (1998).
3   This issue of collaboration between firms raises a separate issue beyond the scope of

Table 2A.1 CIS 3 responses and all UK enterprises

Region	All enterprises		CIS 3 responses		Difference (over-representations)[28] $(d) - (b)$	Error margin $\frac{(e)}{(b)}100$	
	(a)	(b)	(c)	(d)	(e)	(f)	(g)
	No.	Per cent	No.	Per cent	Percentage points	Per cent	Per cent
England	110,308	87.0	6,826	83.5	−3.5	−4.0	±1
East Mids	9,952	7.9	699	8.6	0.7	9.0	±4
Eastern	12,444	9.8	750	9.2	−0.6	−6.5	±3
London	18,804	14.8	974	11.9	−2.9	−19.6	±3
North East	3,846	3.0	444	5.4	2.4	79.1	±4
North West	13,813	10.9	841	10.3	−0.6	−5.5	±3
South East	18,214	14.4	1,012	12.4	−2.0	−13.8	±3
South West	9,659	7.6	621	7.6	0.0	−0.3	±4
West Mids	12,865	10.1	732	9.0	−1.2	−11.7	±4
Yorks & Humbs	10,710	8.4	753	9.2	0.8	9.1	±3
Northern Ireland	3,890	3.1	162	2.0	−1.1	−35.4	±8
Wales	4,195	3.3	379	4.6	1.3	40.2	±5
Scotland	8,382	6.6	805	9.9	3.2	49.0	±3
UK	**126,775**	**100.0**	**8,172**	**100.0**	**0.0**	**0.0**	**±1**

Source: own calculations from UK CIS3 survey

this chapter, namely the question of when collaboration becomes collusion, and how this is (and should be) handled in the context of competition policy. Oughton and Whittam (1996) discuss the relation between co-operation between firms on the one hand and competition policy on the other, combined with an analysis of the benefits to be had from reaping internal and external economies of scale. Co-operative external economies of scale enable small and medium-sized enterprises to pool fixed costs which can result not only in greater efficiency but also, by overcoming entry barriers, thereby increase competition. Thus public sponsorship of such co-operative industrial activities should not be seen as necessarily at odds with promoting competition.

4   See Freeman (1982).
5   For a detailed review of research into technology transfer and public policy, see Bozeman (2000), and for a survey of the literature, see von Tunzelmann (2001).
6   For research specifically on university technology transfer, see the various papers in *the Journal of Technology Transfer Symposium* on 'Economic and Managerial Implications of University Technology Transfer', 2003, Volume 28.
7   The nature and significance of such collaboration is analysed in detail by Coombs *et al.* (1996).
8   For further discussion of the role of trust see Lazaric and Lorenz (1998) and the March 1997 Special Issue of the *Cambridge Journal of Economics* on Contracts and Competition.
9   A point taken up by Dore (1998).
10   The data used by Kitson *et al.* were all for small and medium-sized enterprises. There is a mass of evidence to suggest that collaboration between firms of roughly comparable size tends to be of a very different nature from that between large and small

firms where the power relations are quite different. See the discussion by Oliver and Blakeborough (1998).

11   The importance of this factor is discussed and stressed for example by Pavitt (1993).

12   This is discussed in the case of biotechnology by Orsenigo (1993).

13   Details of the sample are provided in the Appendix.

14   This is consistent with the findings of the 2003 UK Global Enterprise Monitor (GEM) (Harding, 2003) and the 2003 GEM 'High Growth Potential Businesses Specialist Summary', which found 'a weakness in university-industry links that is leading to fewer high potential businesses working with the university sector' (Cowling and Harding, 2003).

15   Chi-squared test.

16   Another reason for the lack of co-operation may lie in the short-termism that prevails in many firms and industries, and a financial system more geared to quick pay-back periods and a high priority to maintaining dividend payout levels than to long-term investment commitments.

17   For a discussion of the policy implications, see also Michie and Oughton (2001).

18   This was from the project reported and discussed in detail by Lyons and Mehta (1997a, 1997b). For a discussion of how firms can establish trust, see Bibb and Kourdi (2004).

19   As argued in detail for example by Tidd *et al.* (1997).

20   On which, see for example the various authors in Archibugi and Lundvall (2001) and the extensive references given and discussed therein. Rosenberg (1982) is an earlier discussion of the role of learning in technological change and diffusion.

21   As stressed by Porter (2003) in his consideration of UK competitiveness policies, and Howells (2011).

22   Using data from CIS3 it is possible to analyse firm co-operation by geographic location – we find that firms that co-operate frequently co-operate at all levels: locally, nationally and internationally.

23   The Regional Development Agencies were abolished by the 2010–15 Coalition Government.

24   CIS 3 includes: C (mining and quarrying), D (all manufacturing), E (electricity, gas and water supply), F (construction), G (from section G only SIC 51 – wholesale trade except of motor vehicles – is included; excluded are wholesale and retail trade, repair of motor vehicles and personal and household goods), I (transport, storage and communication), J (financial intermediation) and K (real estate, renting and business activities).

25   In terms of larger firms, an enterprise is usually the business unit, which must be a legal entity and have a certain degree of autonomy; for smaller firms it is often the whole company.

26   The survey was stratified by Government Office Region in England and by Scotland, Wales and Northern Ireland. Each of these regions contained 12 SIC groups and 5 employment size-bands: 10–49, 50–249, 250–499, 500–999 and 1,000 and more employees.

27   Questionnaires were sent to 19,602 enterprises, approximately 15 per cent of all enterprises on the IDBR. Of these 19,602 firms, 8,172 returned valid answers. This gives a response rate of around 42%.

28   A negative sign means under-representation.

29   The sampling error (SE) for each CIS stratum is calculated as follows:

$$SE = \sqrt{\left(1 - \frac{n}{N}\right) * \frac{\pi(1-\pi)z^2}{n}}$$

where $z = 1.96$, n = sample size, N = population size, and π the proportion in the population with some particular attribute = 0.5.

References

Ahuja, G., 2000, The duality of collaboration: inducements and opportunities in the formation of interfirm linkages, *Strategic Management Journal*, 21, 317–43.

Angrist, J. D. and A. B. Krueger, 2001, Instrumental variables and the search for identification: from supply and demand to natural experiments, *Journal of Economic Perspectives*, 15:4, 69–85.

Archibugi, D. and B.-Å. Lundvall (eds), 2001, *The Globalizing Learning Economy*, Oxford University Press, Oxford.

Bibb, S. and J. Kourdi, 2004, *Trust Matters: For Organisational and Personal Success*, Palgrave Macmillan, London.

Bozeman, B., 2000, Technology transfer and public policy: a review of research and theory, *Research Policy*, 29, 627–55.

Christopherson, C., J. Michie and P. Tyler, 2010, Regional resilience: theoretical and empirical perspectives, *Cambridge Journal of Regions, Economy and Society*, 3, 3–10.

Cohen, W.M. and D. A. Levinthal, 1990, Absorptive capacity: a new perspective on learning and innovation, *Administrative Science Quarterly*, 35, 128–52.

Cohen, W.M. and D. A. Levinthal, 1994, Fortune favors the prepared firm, *Management Science*, 40:2, 227–51.

Coombs, R., A. Richards, P. P. Saviotti and V. Walsh, 1996, *Technological Collaboration: The dynamics of cooperation in industrial innovation*, Edward Elgar, Cheltenham.

Cowling, M. and R. Harding, 2003, *High Growth Potential Businesses, UK GEM: Specialist summary*, The Work Foundation, London.

Dore, R., 1983, Goodwill and the spirit of market capitalism, *British Journal of Sociology*, 34:4, 459–82.

Dore, R., 1998, Innovation and corporate structures: USA and Japan, in J. Michie and J. Grieve Smith (eds), *Globalism, Growth and Governance*, Oxford: Oxford University Press.

Doz, Y. L. and G. Hamel, 1998, A*lliance Advantage: The art of creating value through partnering*, Harvard Business School Press, Boston, MA.

Freeman, C., 1982, *The Economics of Industrial Innovation*, Pinter, London and New York.

Friedman, J. and J. Silberman, 2003, University technology transfer: Do incentives, management, and location matter?, *Journal of Technology Transfer*, 28, 17–30.

Gomes-Casseres, B., 1994, Group versus group: how alliance networks compete, *Harvard Business Review*, July–August, 62–74.

Harding, R., 2003, *UK Global Entrepreneurship Monitor*, The Work Foundation, London.

Howells, J., 2011, Innovation and globalisation: a system of innovation perspective, in J. Michie (ed.), *The Handbook of Globalisation*, 2nd edition, Edward Elgar, Cheltenham.

Huggins, R., 2001, Inter-firm network policies and firm performance: evaluating the impact of initiatives in the United Kingdom, *Research Policy*, 30:3, 443–58.

Kitson, M., J. Michie and M. Sheehan, 2003, Markets, competition, cooperation and innovation', in D. Coffey and C. Thornley (eds), *Industrial and Labour Market Policy and Performance*, Routledge, London.

Lambert, R., 2003, *Lambert Review of Business-University Collaboration*, HM Treasury, London.

Lazaric, N. and E. Lorenz, 1998, *Trust and Economic Learning*, Edward Elgar, Cheltenham.

Link, A. N. and D. S. Siegel, 2002, Generating science-based growth: an Econometric analysis of the impact of organizational incentives on the efficiency of university-industry technology transfer', mimeo, Department of Economics, Rensselaer Polytechnic Institute.

Love, J. and S. Roper, 2001, Location and network effects on innovation success: evidence for UK, German and Irish manufacturing plants, *Research Policy*, 30:4, 643–61.

Lyons, B. and J. Mehta, 1997a, Private sector business contracts: the text between the lines, in S. Deakin and J. Michie (eds), *Contracts, Co-operation, and Competition: Studies in economics, management, and law*, Oxford University Press, Oxford.

Lyons, B. and J. Mehta, 1997b, Contracts, opportunism and trust: self-interest and social orientation, *Cambridge Journal of Economics*, 21:2, 239–76.

Michie, J. and C. Oughton, 2001, Regional and industrial policy, *New Economy*, 8:3, September.

Michie, J. and C. Oughton, 2013, *Measuring Diversity in Financial Services Markets: A diversity index*, SOAS, London.

Michie, J. and C. Oughton, 2014, *Corporate Diversity in Financial Services: An updated diversity index*, London: Building Societies Association.

Mora-Valentin, Eva M., Angeles Montoro-Sanchez and Luis A. Guerras-Martina, 2004, Determining factors in the success of R&D cooperative agreements between firms and research organizations, *Research Policy*, 33:1, January, 17–40.

Moss Kanter, R., 1994, Collaborative advantage: the art of alliances, *Harvard Business Review*, July–August, 96–108.

Mowery, D. C. and N. Rosenberg, 1998, *Paths of Innovation: Technological change in 20th-century America*, Cambridge University Press, Cambridge.

Niosi, J. (ed.), 1999, The internationalization of R&D, special issue, *Research Policy*, 28/2–3.

Oliver, N. and M. Blakeborough, 1998, Innovation networks: the view from the inside, in J. Michie and J. Grieve Smith (eds), *Globalization, Growth, and Governance: Creating an innovative economy*, Oxford University Press, Oxford.

Orsenigo, L., 1993, The dynamics of competition in a science-based technology: the case of biotechnology, in D. Foray and C. Freeman (eds), *Technology and the Wealth of Nations: The dynamics of constructed advantage*, Pinter, London and New York.

Oughton, C. and G. Whittam, 1996, Competitiveness, EU industrial strategy and subsidiarity, in P. Devine, Y. Katsoulacos and R. Sugden (eds), *Competitiveness, Subsidiarity and Industrial Policy*, Routledge, London, 58–103.

Oughton, C. and G. Whittam, 1997, Competition and cooperation in the small firm sector, *Scottish Journal of Political Economy*, 44:1, 1–30.

Oughton, C., M. Landabaso and K. Morgan, 2002, The regional innovation paradox: innovation policy and industrial policy, *Journal of Technology Transfer*, 27, 97–110.

Oxley, J. E. and R. C. Sampson, 2004, The scope and governance of international R&D alliances, *Strategic Management Journal*, 25: 723–49.

Pavitt, K., 1993, What do firms learn from basic research?, in D. Foray and C. Freeman (eds), *Technology and the Wealth of Nations: The dynamics of constructed advantage*, Pinter, London and New York.

Porter, M., 1990, *The Competitive Advantage of Nations*, Macmillan, London.

Porter, M., 2003, *UK Competitiveness: Moving to the next stage*, DTI, London.

Rosenberg, N., 1982, Learning by using, *Inside the Black Box: Technology and Economics*, Cambridge University Press, Cambridge.

Siegel, D. S. and B. van Pottelsberghe de la Potterie, 2003, Symposium overview:

economic and managerial implications of university technology transfer (selected papers on university technology transfer from the Applied Econometrics Association Conference on 'Innovations and Intellectual Property: Economic and Managerial Perspectives'), *Journal of Technology Transfer*, 28, 5–8.

Smith, K. G., S. J. Carroll, and S. J. Ashford, 1995, Intra- and interorganizational cooperation: toward a research agenda, *The Academy of Management Journal*, 38:1, 7–23.

Teece, D. J., 1992, Competition, cooperation, and innovation: organizational arrangements for regimes of rapid technological progress, *Journal of Economic Behavior & Organization*, 18:1, 1–25.

Tether, B., 2002, Who cooperates for innovation, and why: an empirical analysis, *Research Policy*, 31, 947–67.

Tether, B. and P. Swann, 2003, Sourcing science: the use by industry of the science base for innovation; evidence from the UK's innovation survey, Cric discussion paper no. 64, August 2003.

Tidd, J., J. Bessant and K. Pavitt, 1997, *Managing Innovation: Integrating technological, market and organizational change*, Wiley, Chichester.

von Tunzelmann, G. N., 2001, Technology transfer, in J. Michie (ed.), *A Reader's Guide to the Social Sciences*, Fitzroy Dearborn and Routledge, London and New York.

Winter, 1987, Knowledge and competence as strategic assets, in D. Teece (ed.), *The Competitive Challenge: Strategies for industrial innovation and renewal*, Ballinger, Cambridge MA, 159–84.

Yates, I. (ed.), 1992, *Innovation, Investment and Survival of the UK Economy*, The Royal Academy of Engineering, London.

3 Global economic crises and local fortunes

The case of Oxfordshire and Buckinghamshire

Rupert Waters and Helen Lawton Smith

1 Introduction

A motivation for regional policy lies in the belief that regional economic disparities are generally regarded as being socially and economically undesirable.
Therefore economically efficient policies need to be devised to reduce regional
disparities in employment and income, especially in times of economic recession.

Studies on the UK North–South divide in the 1980s and 1990s (Martin 1988,
2004; Martin and Tyler 1991, 1994) drew attention to the imbalance between
growth in the south of the UK when compared to the North. Gardiner *et al.* (2012)
argue that, while the problem has traditionally been identified as that of growth
lagging in the North, in more recent recessions attention has been focused on the
rising concentration of economic activity and growth in the South East and
London and the possible adverse impacts on the rest of the country. Regional
disparities in economic performance are now greater than in any other country in
Europe (Martin *et al.* 2015). The policy response has been that of rebalancing the
economy, with particular attention paid to sectoral, public, private and spatial
aspects of imbalances in the UK economy (see also UKCES 2011).

By 2014, UK GDP had returned to pre-recession levels. This chapter reviews
the comparative fortunes of UK sub-regions through this recovery. It reviews the
extent to which rebalancing has occurred by examining patterns of regional
performance. The case studies of Buckinghamshire and Oxfordshire, neighbouring counties in the South East of England, compare local variations in fortunes in
times of global recessions, to better understand the reasons for their strong
economic performance in times of recession.

The two areas are compared because, although they have similar performances
across a range of indicators including productivity, employment rate and occupational profile, their institutional and industrial structures are markedly different,
with Buckinghamshire's business base being strongly orientated to micro business (firms employing fewer than 10 people and especially fewer than five), while
Oxfordshire has an internationally renowned research infrastructure – including
the University of Oxford and several national research laboratories, in addition to
other higher education institutions like Oxford Brookes University.

Previous studies have shown that places grow when they generate new firms and when new industries emerge and grow as clusters evolve (Mohr and Garnsey 2010; Trippl and Tödtling 2008; Fritsch 2014). Other studies have focused on the role of firm-level innovation in sustaining growth (see for example Archibugi *et al.* 2012). Underpinning entrepreneurship-led growth is the supposition that entrepreneurship can be stimulated at the regional level by policy intervention. The dilemma for policy makers remains that entrepreneurial regions occur independently of politics (Fritsch and Storey 2014), with regions with high levels of entrepreneurship and start-up activity expected to continue to experience the same high levels in the future. Fritsch and Storey argue that the persistence of new business formation demonstrates the existence of long lasting place-based cultures of entrepreneurship.

To consider the performance of these two counties, national datasets are used to review the performance of their local economies over the recession compared to the national pattern using gross value added data supported by other sources. This provides evidence on where the effects of the recession have been felt and hence on the relationship between this and the sustainability of the economy based on its successful clusters (Spencer *et al.* 2010). In particular this research focuses on entrepreneurship and innovation as explanatory factors for the capacity of a local economy to adapt and change in the face of economic shocks.

2 Regional imbalances in performance – rationale and measures

2.1 Explaining rebalancing in the UK context

While regional disparities in economic performance are not in dispute, 'rebalancing' means many things. As UKCES (2011) points out, it can refer to the balance between public and private sector, a dependence on too narrow a band of sectors (e.g. finance in the case of the UK), as well as spatially. UKCES also suggests that rebalancing also describes the need to address differences in public spending and receipts, imports and exports, domestic consumption and business investment. More generally it refers to differences between economic, social and environmental outcomes. Hence the nature and scale of rebalancing is complex. This complexity is also matched by lack of precision in the rationale for government to intervene in the rebalancing process and hence vagueness in the conceptual underpinnings for intervention.

With the emergence of the *Northern Powerhouse* in 2015, an alliance of Northern city regions and Local Enterprise Partnerships (LEPS, discussed below) which aims to 'rebalance the country's economy and establish the North as a global powerhouse',[1] it is spatial rebalancing, particularly in England, that has become a most prominent agenda. The continued rise of the services sector compared to manufacturing and production more broadly in the UK from 2008–2015 has had the effect of frustrating sectoral rebalancing.

An explanation for the persistence of spatial economic imbalances in the UK is the distribution of interconnected sources of power. Martin *et al.* (2015) argue that it has resulted from the progressive concentration of economic, political and financial power in London and the greater South East region, including Buckinghamshire and Oxfordshire, giving the UK one of the most centralised systems among OECD countries.

In support of the claim that the Government is 'getting the whole of Britain back to work with a truly national recovery', the Chancellor of the Exchequer noted in his 2015 Budget Speech[2] that in the year to the last quarter of 2014 employment had grown fastest in the North West of England. However, over the recession (from the second quarter of 2008 to the first quarter of 2015), employment in the Greater South East[3] rose 8.3 per cent compared to 4.6 per cent across the UK. Within this, London was the only region to better the national rate, rising 13.9 per cent ahead of the 4.6 and 4.5 per cent, respectively, recorded in the East and South East regions.

2.2   Regional imbalances in theory

Regional disparities in employment and income, especially in times of economic recession, and solutions to overcome them have long interested academics from a variety of disciplines, particularly geographers. Explanations for regional imbalances as suggested in the various definitions of imbalances above reflect this complexity. Research has analysed industrial structures, patterns of entrepreneurship and innovation and the association of performance with localised concentrations of activity – for example agglomerations and clusters – over time, and the role of the state. As UKCES (2011) argues, the analysis of the nature of underlying causes of imbalances has implications for whether government can and should intervene. As Massey (1979) argued, in order for there to be effective regional policies, it is necessary to understand in what sense there is a regional problem.

Theories which attempt to explain regional imbalances date back some 80 years (see Gardiner *et al.* 2012). They conclude that the various theoretical perspectives (such as those that look at sectoral and spatial imbalances (as for example Hirschman 1958); at cumulative causation (Myrdal 1958; Kaldor 1970, 1981); at the uneven and combined regional development economic structure model (Rowthorn 2010); or at increasing returns from spatial agglomeration as found in the new economic geography models (Krugman 1991; Fujita *et al.* 1999) indicate that the tendency to regionally imbalanced growth is inherent in the market process.

However, theories differ in the extent to which there is a trickle down to disadvantaged regions from national growth. What they do have in common is the assertion that industrial structure, especially the presence of competitive export sectors, plays a critical role in the economic performance of a region (see also Massey 1979; DBIS 2015). Moreover, sectorally unbalanced growth and regionally unbalanced growth are both likely to be self-reinforcing.

UKCES (2011, iv) suggest that rationales for policy intervention are most evident within a neoclassical growth framework, but also within more recent thinking about new economic or functional geographies, where spatial differences are likely to persist because of agglomeration economies. Hence endogenous growth and new economic geographies, as identified by Gardiner *et al.* (2012), have become more influential in economic policy: they 'address the under utilisation of resources and improve the competitiveness of places focusing on place-based growth rather than redistribution' (ibid., v).

While innovation was not considered by Gardiner *et al.* (2010), Archibugi *et al.* (2012) identify a number of recent empirical studies that explore firms' innovative behaviour before and during economic recessions in different contexts – countries and regions. They cite Kanerva and Hollanders (2009) who, in analysing Innobarometer data for Europe,[4] find no association between firm size and decline in investment during 2008. Their results suggest that highly innovative firms continue to invest in innovation even during economic downturns. Similarly, Alvarez *et al.* (2010), in their analysis of Chilean manufacturing firms, explore firms' responses to the financial crisis of 1998. They find a positive association between firm size and organisational innovations, but no impact of financial constraints on innovation performance during the crisis. In contrast, Antonioli *et al.* (2010) find that, in their analysis of firms located in Italy's Emilia-Romagna, small and medium-sized enterprises (SMEs) were more innovative during the recent crisis when compared with large companies. Further, younger businesses supplying foreign multinationals or suffering export shocks were more likely to stop innovating. Filippetti and Archibugi (2011) explored firms' innovation investment in Europe and found that (a) the crisis brings about a reduction in the willingness of firms to increase innovation investment but that (b) strong National Systems of Innovation help firms to retain their investment in innovation.

Gardiner *et al.* (2012) also did not consider entrepreneurship as a driving force of economic development, as identified by Schumpeter (1934). Fritsch and Aamoucke (2013) propose that what leads to some places performing better is more entrepreneurship and public sector funding. These authors found a strong relationship between the presence of universities and other types of public research institutes and the emergence of new businesses in innovative industries, consistent with the knowledge spillover of entrepreneurship (see Audretsch and Keilbach 2005; Acs *et al.* 2009). Fritsch and Aamoucke also found that the size of institutions is important but had a smaller impact than sheer presence. Moreover, there was a modest but not significant positive effect for public research institutions in adjacent regions, indicating regional spillovers. They found that the impact was especially significant for start-ups in high technology and technologically advanced manufacturing but not for technologically advanced services. However, their study was based on Germany with its much larger regions than the counties of Buckinghamshire and Oxfordshire.

In summary, the underlying nature of sectoral and spatial imbalances has been found to be related to inherited industrial structures. Within this broad explanation are place-specific factors such as the presence of competitive export sectors, innovating small firms and entrepreneurship, and in some cases public sector

funding for innovation. Moreover, both sectorally unbalanced growth and regionally unbalanced growth are likely to be cumulative and self-reinforcing. Next we
turn to the cases of Oxfordshire and Buckinghamshire to explore these themes.

3   The Oxfordshire and Buckinghamshire economies in times of recession

3.1   *The UK context*

The UK government's 2011 Plan for Growth aimed, 'to achieve strong, sustainable and balanced growth that is more evenly shared across the country and
between industries'. UKCES (2011) finds that the UK's pattern of GDP by
components of expenditure is typical of several other large developed economies
such as Canada, France and Italy, and hence its rebalancing challenges are not
unusual. However, they identify differences in productivity growth and skills
disparities as key elements. This, UKCES argues, provides a strong argument for
interventions designed to assist in the rebalancing process.

Gardiner *et al.* (2012) noted that the south of the country began to pull away
from the North during the 1980s, creating the so-called 'North-South divide'
(Martin 1988, 2004; Martin and Tyler 1991, 1994). The gap widened in the 1990s
with London gaining most throughout 1992–2007. It was 'London that powered
the growth of the economy of the South, and to a large extent the national economy as a whole' (Gardiner *et al*: 16). They also found that the effects of the
industry mix and region-specific elements are clear. The North's adverse industrial structure and a relative lack of competitiveness in its industries contributed
to the region's poorer growth performance relative to the national economy. This
got worse during the 1990s. London's economy got better, especially with rapid
growth in the financial and business services sector. However, the declining
manufacturing sector now accounts for only 8 per cent of jobs nationally,
although the North continues to be dependent on it. Manufacturing's share of UK
economic output (in terms of gross value added – GVA) has been in steady
decline for many decades, from more than 30 per cent in the early 1970s to 10 per
cent in 2014. This, however, is a reflection of gains made by other industries,
particularly the services sector, rather than significant falls in manufacturing
output (Rhodes 2015). Both these patterns have meant that national growth was
spatially skewed.

Gardiner *et al.* (2012) find no evidence for the view that faster growth at
national level leads to a reduction in spatial imbalance in UK – in fact, it is the
opposite. According to Massey (1979), structural factors are important influences
on the degree of regional imbalance. They conclude that the UK is an example of
a significant change in structure of a national economy over a short period of time
having a dramatic impact on the degree of spatial imbalance.

These cumulative effects present a challenge for policy. In the UK, the political
context is the government's view published in 2015[5] that the UK needs to make
significant improvements to productivity across the regions; and the government

is committed to further radical devolution of power within England. This would give local leaders more opportunity to drive efficiencies by bringing budgets and powers closer to the point of use. It would also improve outcomes through giving local people greater influence over how services are delivered. The 2015 Spending Review was set to establish how spending can be used to rebalance the economy, including by building a Northern Powerhouse. As part of the Spending Review, the government is proposing to look at transforming the approach to local government financing and further decentralising power, in order to maximise efficiency, local economic growth and the integration of public services.

Prior to the Spending Review, the Local Growth Fund, which was recommended by Lord Heseltine in his 2013 report on UK growth[6] was designed to be an important part of the government's commitment to empower local places by giving them the tools they need to drive economic growth. The 2013 Autumn Statement confirmed that central government departments would devolve at least £12 billion from 2015–2016 to 2020–2021 to the Local Growth Fund. As part of the Spending Review process, the government is intending to identify which budgets will be devolved into the Local Growth Fund to support economic development across the country. This would give more funding to local communities for their priority projects.

The 2015 Spending Review is a continuation of the UK's centralised policy system which gives limited autonomy to the regions/localities. The Coalition government (2010–2015) restructured the sub-national policy system by removing the regional tier of economic governance to one focused on localities, often counties or in some cases groupings of local authority units including counties.

Pickles and Cable (2010) stated,

> in its regional structures the last government bequeathed a cumbersome and undemocratic bureaucracy that is unfit for the task. Regional Development Agencies focused on bidding for and spending Whitehall money. They are too cumbersome, costly and unrepresentative to get us through the current economic crisis. A country arbitrarily divided into unnatural blocks such as the 'South-West' and the 'East Midlands' runs against the economic grain. But nor can ministers rebalance economies as diverse as those of Leeds, Liverpool and Tees Valley from our offices in Whitehall.

So a commitment is given to sub-national economic intervention but not to regional development agencies (RDAs).[7]

However, the South East England Development Agency (SEEDA), which included the counties of Buckinghamshire and Oxfordshire, was along with the other RDAs positively evaluated in a report by PriceWaterhouseCoopers for BERR (2009). They were replaced by Local Enterprise Partnerships (LEPS). Despite not being initially intended to necessarily cover the whole country, all of England is now covered by at least one LEP.

The LEPs' responsibilities grew markedly over the course of the 2010–2015 Parliament. In the Heseltine Review it was argued that, while LEPs were 'the

Government's chosen engine of local growth' they 'simply do not, though, have the authority or resource to transform their locality in the way our economy needs'. Of Heseltine's 89 recommendations, 81 were at least partially accepted, including the creation of what became the Local Growth Fund. All 39 LEPs have since received their first instalments of their share of at least £12 billion to be spent in local economies through a series of 'Local Growth Deals'.

Since then there have been two rounds of Local Growth Funding, with a third round of awards to follow. Oxfordshire has been awarded £118.4m and Buckinghamshire £52.0m, to be supplemented by contributions of at least £100m and £28.8m from local partners in Oxfordshire and Buckinghamshire respectively, as well as any of the £126m awarded to the South East Midlands LEP to be spent in the Cherwell (north Oxfordshire) or Aylesbury Vale (north Buckinghamshire) districts.

3.2   Indicators of county-level performance in recessionary times: Oxfordshire and Buckinghamshire

Both counties are prosperous and entrepreneurial. In 2013 gross value added in Oxfordshire and Buckinghamshire stood at £19.2bn and £14.1bn, respectively. Both economies showed comparatively strong performances over the recession, with GVA rising 25.5 per cent in Oxfordshire from 2007 to 2013 and 23.6 per cent in Buckinghamshire, the eighth and eleventh highest rates of growth among the UK's 134 NUTS3 regions and first and third among LEPs. This increase in output was accompanied by strong growth in jobs, with Buckinghamshire ranking behind only London among LEPs, with a 3.7 per cent increase in total workplace-based employment from 2009 to 2013, and Oxfordshire ranking seventh with 2.1 per cent growth, well ahead of the 1.6 per cent recorded across England as a whole. In 2013, only 23 of the 134 NUTS3 regions in the UK recorded productivity, measured by GVA per hour worked, above the national level, with Buckinghamshire and Oxfordshire ranking sixth and fourteenth respectively.

Both local economies are well represented in many of the UK's industrial strategy sectors (DBIS 2012, 2015), which are also mainly innovation intensive. These are aerospace, agricultural technologies, automotive, construction, information economy, international education, life sciences, nuclear, offshore wind, oil and gas, and professional and business services. Oxfordshire has the highest employment-based location quotient in the education sector of any LEP and has an over-representation in the automotive, life science, information economy and professional and business services sectors,[8] while Buckinghamshire is over-represented in life sciences, the information economy, aerospace, education, and professional and business services.

As well as being orientated towards those sectors expected to experience growth, the two economies are over-represented in the most knowledge-intensive occupations when compared to the country as a whole, as shown in Tables 3.1 and 3.2. However, Buckinghamshire's small firm orientation means that it is most strongly over-represented, using the Standard Occupational Code[9] (SOC), for

Table 3.1 Oxfordshire's occupational structure by industry, 2011 (England=100)

	All categories: Industry	A, B, D, E Agriculture, energy and water	C Manufacturing	F Construction	G, I Distribution, hotels and restaurants	H, J Transport and communication	K, L, M, N Financial real estate, professional and administrative activities	O, P, Q Public administration, education and health	R, S, T, U Other
1. Managers, directors and senior officials	106.3	125.2	107.9	100.4	107.0	131.4	99.3	124.1	103.5
2. Professional	125.4	92.4	126.1	84.6	113.4	136.2	135.8	1112	135.6
3. Associate professional and technical	108.3	104.9	118.1	94.1	105.5	120.0	92.3	114.0	100.9
4. Administrative and secretarial	91.2	118.8	103.0	94.5	105.8	100.3	84.3	81.6	101.2
5. Skilled trades	97.2	102.2	89.3	105.4	99.5	72.9	124.6	146.6	123.4
6. Caring, leisure and other service	89.2	170.5	116.4	106.4	97.9	53.6	87.4	78.2	92.8
7. Sales and customar service	84.1	37.1	101.8	96.9	95.4	82.9	65.9	82.2	68.3
8. Process, plant and machine operatives	82.5	91.8	90.9	89.4	92.5	68.3	87.4	119.0	102.5
9. Elementary	88.7	98.1	85.0	99.8	97.5	82.5	84.7	95.8	74.7

Source: Census, ONS, 2011

directors and senior officials (SOC 1); while in Oxfordshire professional occupations (SOC2) are the most over-represented. In manufacture the knowledge-intensity of work in these economies is most marked, with Buckinghamshire and Oxfordshire respectively over-represented by 33.1 and 26.1 per cent for professionals in the workforce.

The technical specialism in the two local economies means that, although GVA per hour worked is high, 21 per cent of businesses in both Buckinghamshire and Oxfordshire acknowledged skills shortages in the 2013 UKCES Employers' Skill Survey,[10] the joint second highest among the 39 LEPs; and employers in both economies find highly skilled workers harder to get than do companies across the country as a whole.

Table 3.2 Buckinghamshire's occupational structure by industry, 2011 (England=100)

	All categories: Industry	A, B, D, E Agriculture, energy and water	C Manufacturing	F Construction	G, I Distribution, hotels and restaurants	H, J Transport and communication	K, L, M, N Financial real estate, professional and administrative activities	O, P, Q Public administration, education and health	R, S, T, U Other
1. Managers, directors and senior officials	125.1	111.4	142.6	126.6	115.1	137.7	121.3	133.2	108.0
2. Professional	103.9	77.0	133.1	99.9	145.7	123.9	99.4	101.1	91.3
3. Associate professional and technical	110.5	110.2	142.9	89.9	145.5	130.7	100.1	97.6	102.0
4. Administrative and secretarial	96.6	131.4	127.2	107.6	120.1	93.6	95.4	83.3	91.4
5. Skilled trades	106.4	104.8	93.6	100.6	101.9	116.8	141.2	101.1	127.9
6. Caring, leisure and other service	100.7	161.2	89.1	64.0	89.1	87.7	92.5	107.8	102.9
7. Sales and customar service	78.4	73.5	118.3	75.7	88.5	63.8	79.7	71.7	83.0
8. Process, plant and machine operatives	74.1	91.4	57.1	83.4	83.6	67.7	84.9	92.5	99.5
9. Elementary	80.0	91.0	57.1	87.7	80.1	71.3	78.5	87.5	84.7

Source: Census, ONS, 2011

Entrepreneurship is strong in Buckinghamshire. In 2013, 86 new businesses for every 10,000 people living in the local authority started trading in Buckinghamshire, the fourth highest rate among LEPs behind London, Hertfordshire and Thames Valley Berkshire, well ahead of the 67.5 recorded across the country as a whole and the 64.9 recorded in Oxfordshire. Oxfordshire's lower rate reflects its larger average size of businesses and the presence of several very large public and university sector employers in the county. In contrast, Buckinghamshire has the lowest share of public sector employment in its labour market of any LEP, at only 13.1 per cent.

Although employment growth over the recession has been strong in Oxfordshire and Buckinghamshire, between the 2001 and 2011 Censuses of

England and Wales the number of people employed in Oxfordshire rose 12.0 per cent, only the fourteenth highest increase among the 39 LEPs and slightly below the 12.1 per cent increase recorded across England, while Buckinghamshire's increase of 7.0 per cent was the third lowest among LEPs. Similarly, although GVA growth has been comparatively strong since the recession, in 2001–2007, Oxfordshire and Buckinghamshire's growth rate ranked them only one-hundred-and-eighth and one-hundred-and-thirty-third, respectively, among the 139 NUTS3 areas in Great Britain.

While these economies have always performed well compared to the rest of the country, rebalancing could be observed. While London was moving ever further from the rest of country, responsible for 19.7 per cent of GVA in 2001, 20.6 per cent in 2007 and 22.2 per cent in 2013, Oxfordshire and Buckinghamshire were being slowly caught by other parts of the UK until the recession, where their resilience has seen them re-establish and extend their lead. For example, together Oxfordshire and Buckinghamshire produced 56.7 per cent of Greater Manchester's GVA in 2001, falling to 53.8 per cent in 2007, before rising again to 59.1 per cent in 2013, having peaked at 59.9 per cent the previous year.

4 Policy responses and policy evolution

By 2013, the Government had identified the action to be taken to allow each locality to reach its potential (HM Treasury, 2013). As well as creating LEPs, and business rate retention, the list of programmes designed to support rebalancing comprised City Deals, Enterprise Zones, the Regional Growth Fund and the Growing Places Fund. The Regional Growth Fund is a £2.6 billion fund operating across England from 2011 to 2016. It supports projects and programmes that are using private sector investment to create economic growth and sustainable employment. The Growing Places Fund provides a further £500m to enable the development of local funds to address infrastructure constraints and promote economic growth.

The policy responses in the two counties to rebalancing have differed because of the timing of institutional structures following the demise of the RDAs in 2010, hence also in their ability to access central government funds. In its last full year of operation, SEEDA, whose territory was home to 8.8m residents and 400,000 businesses, had net expenditure of £117m, having spent more than £1bn in its first seven years of operation from 1999 across the whole of its area. Buckinghamshire featured in SEEDA's plans as forming part of the Milton Keynes and Aylesbury Vale Diamond for Growth and benefited from SEEDA's region-wide programmes, although targeted place-based regeneration was markedly less focused on Buckinghamshire than on other parts of the region. Oxfordshire was in the Thames Valley Diamond for Growth, an area which included other strong economies such as Reading and Southampton.

Buckinghamshire fell out of national economic development programmes from then until December 2011, with the ratification of the Buckinghamshire Thames Valley Local Enterprise Partnership. When the first LEPs were endorsed by Government in October 2010's Local Growth White Paper, the county's bid was not among the 24 initially selected, although it found itself between the single county LEPs of Oxfordshire and Hertfordshire, with London and Thames Valley Berkshire to the south. To the north, the South East Midlands LEP had been formed, which included the Buckinghamshire district of Aylesbury Vale. This left an island of 330,000 residents not covered by the new arrangements. Buckinghamshire's early experience of LEPs and localism has been less positive than in Oxfordshire, which was included in the first round of LEPs.

Oxfordshire has done relatively well out of the new programmes. The county secured one of 24 Enterprise Zones as Science Vale,[11] a wave 2 city deal, a Growth Hub from the Regional Growth Fund and £6m of Growing Places Fund. Buckinghamshire has secured only £4.2m from the Growing Places Fund having not been eligible for a city deal as a rural area and not being able to apply for an enterprise zone, being an area without a LEP at the start of the application process.

Furthermore, Oxford continued to secure Treasury funding for specific projects for innovation due to the excellence of its research base and contribution to the *Eight Great Technologies*[12] that support UK science strengths and business capabilities. For example Oxford University was awarded a satellite centre for the £235m Sir Henry Royce Centre for advanced materials in the 2014 Autumn Statement.[13]

The comparative lack of national and regional policy attention paid to Buckinghamshire has led to it producing a unique set of economic development actors. The vacuum between the abolition of the RDAs and the formation of the Buckinghamshire Thames Valley LEP was filled by ten local entrepreneurs investing their time and money to form Buckinghamshire Business First in the spirit of localism and Big Society. Buckinghamshire Business First was launched following this failure to secure a single county LEP. It was launched as a not-for-profit organisation in April 2011 with the aims of creating and retaining jobs for young people and giving a 'powerful voice' to businesses. In the face of austerity, by working together, these actors have helped Buckinghamshire to benefit from the UK's Conservative and Liberal Democrat coalition government's localism agenda, securing record levels of investment.

The Oxfordshire LEP has a different focus, being focused on innovation and entrepreneurship. The vision, as set out in the Oxford and Oxfordshire City Deal initial award from central government in February 2013, is to '[a]ccelerate the growth of the city region's knowledge-based economy'. The aim is to 'unleash a new wave of innovation-led growth by maximising the area's world-class assets, such as the universities of Oxford and Oxford Brookes, and "big science" facilities such as those at the Harwell Oxford Campus and Innovation Campus'. The entrepreneurial ecosystem is included under the form of an Invest in Growth Hub to help small and medium enterprises to grow through better business support – with a particular focus on supporting innovation.[14] Transport and housing also form part of the vision.

Building on the city deal, the Oxfordshire LEP Strategic Economic Plan[15] set the vision that 'by 2030, Oxfordshire will be recognised as a vibrant, sustainable, inclusive world leading economy, driven by innovation, enterprise and research excellence'. The plan acknowledges that there will have to be a 'strategically astute focus' on the elements identified by the LEP as being of central importance, which include: growing the county's technology clusters; raising the annual rate of new job creation from 0.8 to 1.0 per cent, attracting a minimum of 30 'new high value' international investments each year; building at least 93,560 homes by 2031; £815m of highways infrastructure improvements; and securing an additional 1,150 apprenticeships for young people in growth and priority sectors.

The first Local Growth Deal between Government and the Oxfordshire LEP[16] saw them agree to invest in a range of science-orientated projects as well as more general infrastructure schemes. Science schemes included: the creation of the Centre for Applied Superconductivity to coordinate the interaction between key industry players, Oxford University, cryogenics companies and end users (including SMEs) on the Harwell campus and at the Culham Centre for Fusion Research Campus; and the development of a technology and innovation training centre in Oxford to address skills shortages across engineering, electrical, design and emerging technologies. Infrastructure schemes included a package of junction and local road changes claimed to support growth in the Headington area of Oxford; flood mitigation measures to reduce the risks of damage to homes, businesses and transport connections caused by excessive flooding; and Didcot Station Car Park Expansion, to include the construction of a deck access car park on the existing car park to make Didcot station a key gateway to Science Vale high-tech cluster and the Enterprise Zone.

In Buckinghamshire, the Strategic Economic Plan[17] set the vision 'to create a vibrant, balanced and competitive Buckinghamshire economy', with the LEP's mission defined as being 'to create the conditions that support our business to compete in the Global Race'. The plan sets overarching goals focusing on economic performance and productivity, skills performance and sustainable communities, with targets including the realisation of an additional £319m in GVA over the plan period (2015–2020); the delivery of at least 5,215 apprenticeship starts per annum; and the addition of 2,800 new jobs above the pre-recession peak, while ensuring the youth claimant count is no more than three times the overall claimant count rate.

The Local Growth Deal[18] between Buckinghamshire Thames Valley LEP and the Government focused on five key priorities, which together would deliver 'at least 4,000 jobs and allow 600 homes to be built': improving north–south transport connectivity within the region to accelerate the delivery of key housing and employment sites; improving the integration of transport links to Crossrail and East West Rail stations; investing in skills infrastructure and programmes to tackle welfare dependency and upskill young people; creating high quality private sector jobs by supporting businesses to grow; and addressing housing affordability and supply. Specifically the deal included a commitment from Government to invest £12m in the Aylesbury Eastern Link Road; £8.5m in the High Wycombe and Southern Quadrant scheme; and £6.1m in the A355 Improvement Scheme to improve north–south connectivity. And £1.8m is to be invested in improving bus and cycle connections between Buckingham and the East West Rail station in Winslow, together with £1.5m to improve sustainable transport links to the Crossrail station in Taplow (linking Slough and Maidenhead) to improve the integration of transport links to CrossRail and East West Rail.

The emphasis on transport and infrastructure in the first local growth deal reflects the reliance on devolution of the Department for Transport's budgets, with other departments expected to see greater devolution of their spending in subsequent rounds of the Local Growth Fund. This change of emphasis of the sources of funding started to be felt in January 2015's second tranche of Local Growth Fund agreements, which in Buckinghamshire were to allow the provision of access to finance to small and medium-sized businesses, so that they can grow and create new jobs; the extension of superfast broadband across Buckinghamshire to areas of high business population; and the provision of business incubation and innovation centres for small and medium-sized businesses and start-ups, as well as further transport spending, including improved rail links and facilities between Marlow and Maidenhead.

In Oxfordshire, the expansion[19] included a package to improve transport in North Oxford and enable the Northern Gateway Development, which will provide business and research space, and new homes; establishing the Activate Care Suite to improve adult social care and healthcare in Oxfordshire; and transport and site improvements to support the 'Oxpens development' (a site situated between Oxford railway station, Oxford's Westgate shopping centre and the river Thames), which will provide much needed office and research space and new homes in the heart of Oxford.

Ahead of the third round of the Local Growth Fund and in response to the devolution offer set out in the Plan for Productivity that accompanied 2015's Summer Budget, Oxfordshire and Buckinghamshire formed a strategic alliance with a third and adjacent county, Northamptonshire. This third partner, while less advantaged in terms of its industrial structure than either Oxfordshire or Buckinghamshire, is nonetheless advantageously positioned at the heart of Britain's transport networks. Calling itself 'England's Economic Heartland', this alliance submitted its Strategic Transport and Infrastructure Proposition[20] to the Chancellor of the Exchequer in July 2015. This sets out how the County Councils and the LEPs will work together in order to grow their economies by 20 per cent to 2020. The argument is that, by working collectively, they can reduce the costs of infrastructure projects by up to 40 per cent and accelerate delivery by a third. The three have also been in contact with neighbouring unitary authorities in order to develop the alliance further.[21]

5   Conclusions

The theme of this chapter is the persistence of spatial and sectoral inequalities in the economic fortunes of UK regions. The policy response has been that of rebalancing the economy, in the form of trying to make the weaker regions stronger and catch up with the stronger regions. Since the 1980s, London and the South

East, which includes the counties of Buckinghamshire and Oxfordshire, have outperformed the rest of the country as measured by GVA. Even in recessionary times, counties such as Oxfordshire and Buckinghamshire have been shown to be more resilient to global economic crises and recessions than have areas in the north of the country, which have adverse industrial structures and less competitive industries.

Explanations for the source and persistence of regional imbalances have drawn on research which has analysed industrial structures, patterns of entrepreneurship and innovation, the association of performance with localised concentrations of activity for example agglomerations and clusters over time, and the role of the state.

What the cases of Buckinghamshire and Oxfordshire show is that different combinations of factors identified in these complementary theories produce similar high economic performance. Entrepreneurship and self-employment have been important factors in the UK labour market's unexpectedly strong performance through the recent recession. What the counties have in common is that they have favourable industrial structures, with economies that have clusters of knowledge-intensive sectors, albeit many of them being different. This is reflected in the profile of their workforces with higher representations of skills at SOC 1 (directors and senior officials) in Buckinghamshire and SOC 2 (professionals) in Oxfordshire. Both have very high skill bases and both are entrepreneurial. This is particularly the case with Buckinghamshire, whereas Oxfordshire is shown to be slightly less so. This in part reflects differences in industrial structures. In Oxfordshire, the size of businesses is on average high as a result of the presence of several very large public and university sector employers.

It is in the role of the state and policy responses where the counties differ most. What these differences show is the unequal distribution of interconnected sources of power, even within the UK's centralised system of government (Martin *et al.* 2015). While Oxfordshire has been a substantial beneficiary of funds under decentralisation policies introduced from 2010, and therefore able to use funding to support its priorities, particularly for innovation, Buckinghamshire got Local Growth Funding but pretty much nothing else. Instead, it went for an alternative and locally organised strategy, one which has been shown to be very effective in bringing in investment. Overall, central government investment in local economic development in the two counties has been slight by national standards for similar sized economies, but large by historic standards – perhaps that is what has been delivering results. Either way, both counties are very successful, being top performers amongst the LEPs.

Decentralisation of policies has had further rescaling effects, with the strategic alliance of Oxfordshire, Buckinghamshire and Northamptonshire[22] demonstrating that a larger regional effort, however organised, is also a necessary place for policy responses for sustaining economic development. This brings together a county (Northamptonshire) that has had an adverse industrial structure, but which is located at the centre of Britain's transport networks, with two high-growth economies.

Notes

1   www.gov.uk/government/uploads/system/uploads/attachment_data/file/427339/the-northern-powerhouse-tagged.pdf (accessed August 7 2015).
2   www.gov.uk/government/speeches/chancellor-george-osbornes-budget-2015-speech (last accessed August 6 2015).
3   The three NUTS1 regions of London, South East England and the East of England.
4   Part of Eurobarometer (EC) data for Europe, based on annual polls of businesses and the public.
5   www.gov.uk/government/uploads/system/uploads/attachment_data/file/447101/a_country_that_lives_within_its_means.pdf (accessed July 21 2015).
6   www.gov.uk/government/publications/no-stone-unturned-in-pursuit-of-growth (accessed August 7 2015).
7   www.gov.uk/government/speeches/economy-needs-local-remedies-not-regional-prescription (accessed June 23 2015).
8   As defined in the 2013 Witty Review.
9   www.gov.uk/government/uploads/system/uploads/attachment_data/file/423732/codes_of_practice_april_2015.pdf (accessed August 9 2015).
10 www.gov.uk/government/uploads/system/uploads/attachment_data/file/327492/evidence-report-81-ukces-employer-skills-survey-13-full-report-final.pdf (accessed August 9 2015).
11 www.sciencevale.com (accessed August 9 2015).
12 See for example: www.gov.uk/government/publications/eight-great-technologies-infographics (last accessed August 6 2015).
13 www.gov.uk/government/uploads/system/uploads/attachment_data/file/382328/44695_Autumn_Statement__Print_ready_.pdf (last accessed August 6 2015).
14 www.gov.uk/government/uploads/system/uploads/attachment_data/file/276205/Oxford-Oxfordshire-City-Deal.pdf (accessed November 17 2014).
15 www.oxfordshirelep.com/sites/default/files/Oxford%20SEP_FINAL_March14_0.pdf (last accessed August 10 2015).
16 www.gov.uk/government/uploads/system/uploads/attachment_data/file/398871/27_Oxfordshire_Growth_Deal.pdf (last accessed August 10 2015).
17 www.buckstvlep.co.uk/uploads/downloads/BTVLEP-SEP-FINAL-02-04-14.pdf (last accessed August 10 2015).
18 www.buckstvlep.co.uk/uploads/downloads/BTVLEP-SEP-FINAL-02-04-14.pdf (last accessed August 10 2015).
19 www.gov.uk/government/uploads/system/uploads/attachment_data/file/399432/Oxfordshire_Factsheet.pdf (last accessed August 10 2015).
20 www.buckstvlep.co.uk/uploads/downloads/Tri-County%20Brochure.pdf (last accessed August 10 2015).
21 www.oxfordshire.gov.uk/cms/news/2015/aug/englands-economic-heartland-fighting-better-transport-links-unlock-future-economic (accessed August 9 2015).
22 www.englandseconomicheartland.com/Pages/home.aspx (last accessed January 22 2016).

References

Alvarez, R., Bravo-Ortega, C. and Navarro, L. (2010) *Innovation, R&D Investment and Productivity in Chile* www.iadb.org/res/publications/pubfiles/pubIDB-WP-190.pdf (accessed August 10 2015).

Antonioli, D., Mazzantim M. and Pini, P. (2010) 'Productivity, Innovation Strategies and Industrial Relations in SMEs. Empirical Evidence for a Local Production System in Northern Italy', *International Review of Applied Economics*, 24, 4, 453–482.

Archibugi, D., Filipetti, A. and Frenz, M. (2012) 'The Impact of the Economic Crisis on Innovation: Evidence from Europe' www.bbk.ac.uk/innovation/publications/docs/Archibugi-et-al-Innobarometer.pdf (accessed August 9 2015).

Audretsch, D. and Keilbach, M. (2005) 'The Knowledge Spillover Theory of Entrepreneurship' web.mit.edu/iandeseminar/Papers/Fall2005/audretschkeilbach.pdf (accessed May 19 2011).

Department of Business, Innovation and Skills (DBIS) (2012) *Industrial Strategy: UK Sector Analysis* www.gov.uk/government/publications/industrial-strategy-uk-sector-analysis (accessed August 10 2015).

Department of Business, Innovation and Skills (DBIS) (2015) *2010 to 2015 Government Policy: Industrial Strategy* www.gov.uk/government/publications/2010-to-2015-government-policy-industrial-strategy/2010-to-2015-government-policy-industrial-strategy (accessed November 23 2015).

Filippetti, A. and Archibugi, D. (2011) 'Innovation in Times of Crisis: National Systems of Innovation, Structure and Demand', *Research Policy*, 40, 2, 179–192.

Fritsch, M. (2014) 'New Firm Formation and Sustainable Regional Economic Development – Relevance, Empirical Evidence, Policies', ICER Working Paper Series on Entrepreneurship and Innovation, WP 8, Essex Business School.

Fritsch, M. and Aamoucke, R. (2013) 'Regional Public Research, Higher Education, and Innovative Start-ups: An Empirical Investigation', *Small Business Economics*, 41, 865–885.

Fritsch, M. and Storey, D. (2014) 'Entrepreneurship in a Regional Context – Historical Roots and Recent Developments', *Regional Studies*, 48, 939–954.

Fujita, M., Krugman, P.P. and Venables, A.J. (1999) *The Spatial Economy: Cities, Regions and International Trade*, Cambridge, MA: MIT Press.

Gardiner, B., Martin, R. and Tyler, P. (2012) 'Spatially Unbalanced Growth in the British Economy', working paper CGER No. 1 www.landecon.cam.ac.uk/pdf-files/cv-etc/pete-tyler/cgerworkingpaperno1v5.pdf (accessed March 16 2014).

Hirschman, A. (1958) *The Strategy of Economic Development*, New Haven, CT: Yale University.

HM Treasury (2013) *Investing in Britain's Future*. Cmnd 8669. London: HM Treasury.

Kaldor, N. (1970) 'The Case for Regional Policies', *Scottish Journal of Political Economy*, 18, 337–348.

Kaldor, N. (1981) 'The Role of Increasing Returns, Technical Progress and Cumulative Causation in the Theory of International Trade and Economic Growth', *Économic Appliquée*, 34, 593–617.

Kanerva, M. and Hollanders, H. (2009) 'The Impact of the Economic Crisis on Innovation. Analysis Based on the Innobarometer 2009 Survey', INNO Metrics Thematic Paper, Brussels: European Commission, DG Enterprise.

Krugman, P. (1991) *Geography and Trade*, Cambridge, MA: MIT Press.

Martin, R. (1988) 'The Political Economy of Britain's North–SouthDivide', *Transactions of the Institute of British Geographers*, 13, 389–418.

Martin, R.L. (2004) 'The Contemporary Debate over the North–South Divide: Images, and Realities of Regional Inequality in Late-Twentieth Century Britain', in Baker, A.R.H. and Billinge, M.D. (eds) *Geographies of England: The NorthSouth Divide – Imagined and Material*, Cambridge: Cambridge University Press, pp. 15–43.

Martin, R.L. and Tyler, P. (1991) 'The Regional Legacy of the Thatcher Years', in Michie, J. (ed.) *The Economic Legacy, 1979–1992*, London: Academic Press, pp. 140–167.

Martin, R.L. and Tyler, P. (1994) 'Real Wage Rigidity at the Local Level in Great Britain',

Regional Studies, 28, 833–842.

Martin R., Pike, A., Tyler, P. and Gardiner, B. (2015) *Spatially Rebalancing the UK Economy: The Need for a New Policy Model* www.regionalstudies.org/uploads/documents/SRTUKE_v16_PRINT.pdf.

Massey, D. (1979) 'In What Sense a Regional Problem', *Regional Studies*, 13, 233–243.

Mohr, V. and Garnsey, E. (2010) 'Exploring the Constituents of Growth in a Technology Cluster: Evidence from Cambridge, UK' Centre for Technology Management (CTM) Working Paper 2010/01 http://papers.ssrn.com/sol3/papers.cfm?abstract_id=1923065 (downloaded March 22 2013).

Myrdal, G. (1958) *Economic Theory and Underdeveloped Regions*, London: Methuen.

ONS (2013) 2011 Census, table WP6604EW www.nomisweb.co.uk (accessed 23 November 2015).

Pickles, E. and Cable, V. (2010) www.gov.uk/government/news/a-recovering-economy-requires-local-remedies.

PriceWaterhouseCoopers (2009) *Impact of RDA Spending –National Report – Volume 1 – Main Report*, a report for Department for Business, Enterprise and Regulatory Reform www.bis.gov.uk/files/file50735.pdf (accessed October 26 2012).

Rhodes, C. (2015) 'Manufacturing: Statistics and Policy', House of Commons Briefing Paper No. 01942, August 6 2015.

Rowthorn, R.E. (2010) 'Combined and Uneven Development: Reflections on the North–South Divide', *Spatial Economic Analysis*, 5, 355–362.

Schumpeter, J.A. (1934) [1911], *The Theory of Economic Development*, Cambridge: Harvard University Press.

Spencer, G.M. Vinodrai, T., Gertler, M. and Wolfe, D. (2010) 'Do Clusters Make a Difference? Defining and Assessing their Economic Performance', *Regional Studies*, 44, 6, 697–715.

Trippl, M. and Tödtling, F. (2008) 'Cluster Renewal in Old Industrial Regions – Continuity or Radical Change?', in C. Karlsson (ed.) *Handbook of Research on Cluster Theory*, Cheltenham: Edward Elgar.

UKCES (2011) *Rebalancing the Economy Sectorally and Spatially: An Evidence Review. Volume 1 – Main Report* webarchive.nationalarchives.gov.uk/20140320161942/http:/www.ukces.org.uk/publications/er33-rebalancing-the-economy (accessed August 9 2015).

Witty, Sir Andrew (2013) *Encouraging a British Invention Revolution: Sir Andrew Witty's Review of Universities and Growth* www.gov.uk/government/uploads/system/uploads/attachment_data/file/249720/bis-13-1241-encouraging-a-british-invention-revolution-andrew-witty-review-R1.pdf (accessed November 27 2015).

4 Regional capabilities and industrial resiliency

Specialization and diversification dynamics in Lowell, Massachusetts

Michael H. Best[1]

> For the pattern is more than the sum of the threads; it has its own symbolic design of which the threads know nothing.
>
> Arthur Koestler

Specialization versus diversification

Why are some regions more innovative than others? Is innovation fostered by specialization or diversity? Despite decades of debate, there is no unequivocal answer to these questions. The so-called Marshall-Arrow-Romer view is that innovation is stimulated by local economic specialization where it is driven by knowledge spillovers among local firms in the same or closely related industries. In the view associated with Jane Jacobs, however, innovation is advanced by local economic diversity and heterogeneity as variety increases the scope for interaction, serendipity, and new ideas.

Empirical tests have not settled the matter. Glaeser and colleagues have found support for positive effects on growth for urban industrial diversity (1992). Henderson, in contrast, finds strong support for same-industry concentration on productivity at the county level (1997, 2003).[2]

Michael Porter weighed in on the side of specialization but of clusters of linked industries rather than industries *per se* (2003: 562). He critiques macroeconomic debates on growth and policymaking for being conducted at too high a level of aggregation, but he argues that industry is too narrow a unit because of "externalities across related industries within clusters" (2003: 562). His concept of cluster cuts across sector lines to include suppliers in other sectors as well as service providers, specialized labor inputs, and demanding customers.[3]

Most empirical contributions to the debate share a common research methodology: specialization is defined and measured in terms of the official SIC (Standard Industry Classification) or the new NAICS (North American Industrial Classification System) codes. Using either industry sector or linked sectors as the unit of analysis and measurement has a major drawback: they do not control for differences in business organization. Dynamics of specialization and diversity that operate at the level of the firm and inter-firm relationships may be too finely grained to be captured at the sector level.

Failure to control for business organization is particularly problematic in regional comparisons.[4] For example, the large performance gap between Silicon Valley and Route 128 in the computer industry cannot be traced to differences in specialization measured by employment by SIC or NAICS codes. The SIC sector-defined measures obscured the critical difference in business models in the two regions. The Route 128 minicomputer companies were vertically integrated; the Silicon Valley minicomputer and PC companies were vertically specialized into "open-system," focus and network business models (Grove 1996).[5] Thus, similar measures for output and employment in the two regions masked paradigm differences in firm organization and inter-firm adjustment processes.

The failure to account for inter-firm specialization dynamics is not a problem for high-tech industries alone. The networked groups of vertically specialized small firms that populate the furniture industry of regions in Italy dominated the vertically integrated furniture firms of North London. A regional industry that employed over 16,000 in the 1960s, North London furniture industry employed only 500 by the mid-1980s (Best 1990: 228).

The term networking covers very different forms of inter-firm relations. It can mean simply the commercial business-to-business relations of specialists along the supply chain. It can involve long-term partnering in complementary capability development across numerous new product generations, as in the case of the PC, aircraft (Prencipe 2000), or furniture industries. It can refer to groups of firms cumulatively and jointly advancing distinctive production capabilities, such as precision instruments, or technologies, such as biotech.

Pressure to introduce business organization into economic geography is coming from another direction. Advocates of science and technology policy argue that business organization has important implications for industrial innovation (Broers 2005). The corporate practice of owning research labs was based on the presupposition of a one-way relationship from scientific research to technological advance to market leadership. The presupposition collapsed with the emergence of product-led business models organized to compete on the basis of leadership in the introduction of new products. In the new model, basic research became subservient to technological research, and both to product development.[6]

The new business model has both organizational and geographical implications. Whereas the default organizational context of the science-push model of innovation was the multi-divisional enterprise, today it has become one of a regional innovation system exemplified by Silicon Valley.

However, despite the changes in the real world, business organization remains a shadowy figure in the major theoretical perspectives on economic geography. This is changing. Porter finds that vigorous domestic business rivalry and the "creation and persistence of competitive advantage in an industry" are strongly associated empirically (1990: 117). He adds that "the important influences on ... innovation are intensely local" (1990: 144) and that "competitive advantage is created and sustained through a highly localized process" (1990: 19). The challenge is explicate the specific localized processes that influence innovation and create competitive advantage. Instead, all too often, innovation is conceptualized as an externality, usually in the form of a "knowledge spillover" across co-located firms.[7]

The exception is evolutionary economic geography in which recent contributions have transcended explanations in terms of externalities and gone inside the firm to account for patterns of spatial agglomeration. It has much in common with the capabilities perspective which informs this paper. Both suggest an interpretation of regional specialization that focuses attention on processes of differentiation and specialization within and amongst firms. We start with a review of the contributions of evolutionary economic geography before turning to the capabilities and innovation perspective of Edith Penrose (1995).

From routines to regional capabilities

In a recent survey, Ron Martin (2005: 17) states the following:

> An evolutionary perspective emphasizes dynamic competitive advantage, and the adaptive capabilities of a regional economy to respond to shifts and changes in markets, the rise of competitors, and the development of new technologies. A region's competitive advantage is the complex outcome of its past development—path dependence—and its capacity to create new pathways of development.

But much research remains to be done. Martin adds:

> we know little about the processes of regional economic adaptation, or about why some regions seem to be more adaptive than others .... An evolutionary perspective ... would place ... emphasis on a region's propensity to innovate, both within and amongst firms, and within and amongst its institutions.
>
> (2005: 30)

To address the challenge, "routine," a core concept first developed to explain business success by Nelson and Winter (1982),[8] has been extended to explanations of regional specialization.[9] In the words of Boschma and Frenken (2006: 6): "The emergence of spatial agglomerations is ... to be analyzed in terms of historically grown spatial concentration of knowledge residing in organizational routines." The primary organization is the business enterprise and the selection process for routines is organizational competition: the "starting point is to view organizations as competing on the basis of their routines ... built up over time" (Boschma and Frenken 2006: 6).

The concept of routine in evolutionary economics serves two functions towards meeting the challenge posed by Martin. The first is to shift the concept of competition from price to organization. Whereas conventional microeconomic theory assumes firms make the same product and compete on price, evolutionary theory holds that firms compete on the basis of differentiated routines. It is a short step from competition over routines to Schumpeter's (and Broers's) notion of innovation as new product development:

> as organizations compete on the basis of their routines, competition is driven by Schumpeterian innovation based on new products and technologies requiring new routines rather than on production costs alone as assumed in neoclassical models.
>
> (Boschma and Frenken 2006: 6)

The second function of routine, exemplified by the term "organizational" skills, is ontological, to do with the assumed constitution of economic reality. Production in real companies and economies has irreducible collective and interactive dimensions that are obscured by theories which assume production can be decomposed into independent "factors of production." The skills term in "organizational skills" points to accumulated learning effects and tacit knowledge embedded in production processes which influence organizational productivity and competitiveness.

However, the introduction of routine is less successful in explaining regional specialization. It is not because of lack of effort and it is not without promise. Boschma and Frenken develop the concept "related variety" to distinguish amongst sectors that are closely related from those that are truly diverse: related variety is defined as "complementary capabilities among sectors" (Boschma and Frenken 2006: 13). For example, the previous existence of the bicycle industry within a region gave a regional advantage for the emergence of the auto industry because of components produced by the same or closely related routines. Thus the bicycle and auto industries are an example of "related variety," and greater related variety across sectors in a region, it is argued, is conducive to regional growth.[10]

Related variety offers a different concept of a historically evolving, regionally distinct cluster from the established, uniform cluster concept of Porter. Whereas Porter seeks to identify carbon copy clusters in multiple regions, each constituted by the same linked sectors, the idea of related variety implies small groups of enterprises historically evolving that share common but regionally distinctive "organizational skills" or routines. The concept of related variety is consistent with the empirical observation of "clustered diversity" used by Martin to describe the two most innovative regions in the UK. He observes that the Cambridge "high-tech" cluster is in fact made up of at least seven sub-clusters and that the South East economy, which appears highly diversified, contains several highly specialized clusters of innovative activity (2005: 29).

The application of the concept of routine to economic geography is an important step in integrating business organization into studies of regional economic adaptation. But can routine carry the heavy load to which it has been assigned? It does go inside the "black box" and give distinctiveness and organizational integrity to the business enterprise. But it has a weakness. Routine is a static concept. It cannot explain change in routine, where routines come from, or even what they are.

This critique is not new. Efforts have been made to extend the concept of routine as a description of "stable patterns of organizational behavior" into a theory of organizational change. Zollo and Winter offer the concept of dynamic capabilities "defined as routinized activities directed to the development and adaptation of operating routines" (2002: 1).[11] But the same authors make the candid and surprising comment, given its source: "the literature does not contain any attempt at a straightforward answer to the question of how routines—much less where dynamic capabilities—are generated and evolve" (2002: 9).

On this point we disagree. While Penrose did not use the term, capability, its substance is captured in her concept of growth as "an evolutionary process ... based on the cumulative growth of collective knowledge, in the context of a purposive firm" (1995: xiii). Penrose elaborates a learning firm concept based on a distinction between resources and the productive services of resources. "It is never resources themselves that are the "inputs" into the production process, but only the services that the resources can render."

Resources can be purchased in the market. In contrast, the services that resources render a firm depend upon the distinctive experience and teamwork of its members:

> Experience ... develops an increasing knowledge of the possibilities for action and the ways in which action can be taken by ... the firm. This increase in knowledge not only causes the productive opportunity of a firm to change ... but also contributes to the "uniqueness" of the opportunity of each individual firm.
>
> (Penrose, 1995: 53)

Equally important, a firm is more than a collection of individuals: "it is a collection of individuals who have had experience in working together, for only in this way can 'teamwork' be developed" (Penrose, 1995: 46).

The direction of innovation or knowledge creation in any firm is "closely related to the nature of existing resources (including capital equipment) and to the type and range of productive services they can render" (Penrose, 1995: 84).

She calls attention to economies, not of scale or size, but of growth created by new applications of unused productive services internally generated from previous successful applications. Inducements to expansion "arise largely from the existence of a pool of unused productive services, resources, and special knowledge, all of which will always be found within any firm."

The innovation drive of the Penrosian firm is governed by an iterative dynamic in which firms seek to redeploy unused productive resources to incipient market opportunities and in the process deepen and differentiate their productive services. The learning experience renders yet new or refined productive services, which triggers a new learning and market opportunity detection dynamic.

For terminological simplicity we follow George Richardson (1972) in using the term capability for productive services of resources. Capabilities, like productive services, cannot be developed alone or at once, they are cumulative and collective, they evolve with teamwork and experience, and in the process knowledge is increased. Richardson extended Penrose's internal process of capability development to an inter-firm process in which firms specialize on core capabilities and partner for complementary capabilities.

The focus on organizational capabilities, experience, and skills at the enterprise level has a regional corollary. Production and technological capabilities, for example, are key concepts in this paper. Historically, shifts in regional industrial leadership can be explained in terms of the introduction of plants organized around new, advanced principles of production. Mass production, a capability anchored in the principle of flow, was a revolutionary new means of production that established a step-change in productivity and raised performance standards in cost by an order of magnitude. Equally important, from a regional perspective, the Ford plant was both a demonstration site and a learning center for application of the new principle of production. Engineers, managers, workers, and suppliers simultaneously constructed and learned the new production processes and the complementary skills all of which could be passed on to other firms and other industries.[12] Succeeding generations of engineers, managers and workers were as if plugged into roles within the production system without awareness of its origins or architecture.

While the Ford plant was revolutionary (and ultimately transitory), the principle of flow is enduring.[13] It became a source of regional competitive advantage wherever business enterprises met the challenge of applying the principle to commodity sectors still laboring under pre-mass-production methods. Even today, it remains a means of making a step-change in productivity within factories and regions that have not made the transition but are seeking to engage in high volume production.

However, advances in production capability can be incremental as well as transitional to new principles. A legacy of distinctive production capabilities constructed collectively and cumulatively over many new product generations and diverse applications and diffused into co-located firms is a regional resource that cannot be easily replicated. Most importantly, such capabilities are a platform for a new round of product development initiatives.

Regions have distinctive technological capabilities as well. Seattle and Tacoma, Washington, for example, is a region in which aircraft have been designed and developed for many decades and in which the associated engineering principles and practices are woven into the fabric of industrial life. Individuals plug into long-established routines much as they do into a language community, but their individual skills are as interdependent as the words in the language itself. Both are social as well as individual. They are "deep craft skills" or "organizational skills."

The regional renewal challenge is to either make the transition to more advanced capabilities and/or to convert already established capabilities and skills into new products, new enterprises, and new industrial sub-sectors. The agents of the transition and/or conversion are, in most cases, the business enterprises located in the region. A regional business system which includes networks of specialist enterprises with complementary capabilities is an organizational infrastructure which eases the entry of new enterprises. A new entrant can focus on a core capability and partner for complementary capabilities; a rapidly growing company can tap skills and capabilities no longer required by nearby companies in declining markets.

At the regional level, industrial renewal is a systems integration process in that companies, capabilities and skills are reshuffled, reconfigured, and redesigned to seize market opportunities. In the process, the technological capabilities which were leveraged in the first place are advanced and deepened, thereby reinforcing the region's competitive advantage. In this way technology-driven business enterprises act as vehicles by which regional capabilities are renewed.[14]

Thus a dynamic Penrosian capabilities perspective focuses attention on dynamic specialization processes by which new firms, sectors, "industries," or clusters emerge and evolve. It implies that the long-standing debate has been misguided: specialization and diversification are different parts of a single process. Specialization in capabilities enhances diversity in application, and greater sector diversity enhances market opportunities for new product development which, in turn draws upon and renews specialized regional capabilities. At the same time product differentiation is the leading edge of industrial diversity. It evokes Jacobs's growth perspective of increasing industrial differentiation based on the creation of new sectors.[15]

The challenge is to characterize distinctive regional capabilities. Unlike output and employment, capabilities are intangible. Nevertheless, they are real and, I argue below, subject to indirect characterization methods based on empirical inferences generated by application of a finely granulated technology taxonomy to companies and their products.

Research methodology

The empirical challenge is considerable. The presumption is that regional production and technology capabilities, while intangible, are embedded in the production processes and deep craft skills of a region and manifest in specialized groups of companies and the products that they design and develop. We seek to "discover" the underlying capabilities that, while hidden, impart locational advantage by examining tangible patterns of companies and products.[16] A study of capabilities requires a historical data set of real companies and their products. Official data is no help for two reasons. First, companies are ahistorical and anonymous and second, the classification categories lack the granularity required to capture both differentiation and specialization.

To get inside the companies that populate a region, we have constructed vTHREAD (Techno-Historical Regional Economic Analysis Database) a historical database of approximately 55,000 public and private high-tech producers and their products.[17] The dataset includes the companies and the products they make and, most importantly, is organized in terms of a finely granulated technology taxonomy originally developed by CorpTech. The CorpTech technology product classification system has three major filters. Companies are grouped by 18 primary industry codes and 280 major product codes, which in turn support roughly 3,000 technology-product applications.[18] This degree of granularity is requisite for "discovering" technological capabilities and capturing both special-ization and differentiation processes.

The research methodology is based on the proposition that the existence of a distinctive regional technological capability gives local firms a competitive advan-tage. Therefore, an inter-temporal grouping of companies in closely related technology-product domains is an indicator of an underlying regional capability. The regional capability itself is not observable or directly measurable, but it can be revealed by a study of groups of successful firms with similar technology-product characteristics. Individual firms may enter and exit, but the presumption is that at any point in time there is the critical mass required to sustain the new product devel-opment process and, with it, revitalize the underlying technological capabilities.[19]

Three propositions guide the analysis of industrial renewal presented here.

- The first is a path dependence proposition. A region's production and tech-nology capability legacy is a basis for regional competitive advantage. Today's firms can leverage distinctive capabilities and skills established by preceding companies and in the process advance them for succeeding prod-uct development generations. Here our focus is on inter-temporal evolution of regionally specialized capabilities. The challenge is to characterize these capabilities and their evolution.
- The second is a cluster dynamics proposition. A region's capabilities are also shaped by dynamic, inter-firm, mutual adjustment processes. If the first proposition focused on the dynamics of regional specialization and the evolution of specialized groups, the second focuses on the dynamics of differentiation and transformation within and amongst specialized groups.
- The third proposition is that regional specialization is shaped by both "path dependence" and "cluster dynamics" and that each is better understood within a framework that accounts for both.

The research challenge is to articulate and characterize a range of processes by which the pre-existing production and technological capabilities within a region are leveraged, adapted, integrated, differentiated, and reshuffled. The modus operandi is to first group a region's high-tech enterprises by a finely granulated taxonomy. The companies and the groups are then examined first in terms of capability "path dependence" or inter-temporal evolution of regional specialization and second in terms of inter-firm mutual adjustment processes by which the groups themselves are reconstituted by internal processes of differentiation and transformation.

An industrial laboratory: Lowell, Massachusetts

Lowell, Massachusetts is an interesting location for examining industrial transi-tion and renewal. Industrial renewal occurred not once but twice in recent times. And the two industrial renewals were remarkably different, at least in appearance.

In the beginning, Lowell was a textile city. The population of Lowell grew in synch with the nation's textile industry for nearly a century, expanding from 2,500 in 1826 to over 110,000 in 1920 (Gittell 1995). In the early 1920s, employment in the textile industry accounted for over 40 percent of all manufacturing in Lowell. But between 1924 and 1932, manufacturing employment fell by 50 percent (Flynn 1988: 277). Lowell did not recover from the Great Depression, but employment began to grow in the 1960s primarily in low-wage and declining industries.

The first post-textile era industrial renewal was led by a single company and a single industry. Wang Laboratories, founded in 1951, relocated to the city in 1977. In the peak years of the mid-1980s, Wang Labs employed over 10,000 workers in the Lowell labor market, accounting for approximately 10 percent of total employment and one-third of manufacturing employment in the local economy (O'Connell 1991a). But Wang transformed the industrial landscape only to collapse, and double-digit unemployment returned. The Wang era ended in 1994 with the auction of the 15-acre, three-tower, 1.5-million-square-foot-office complex for $525,000, approximately 1 percent of the $55 million cost to build the facility.

Lowell suddenly went from a success story to a recipe for industrial decline. Once more the sources of decline were attributed, properly, to excessive reliance on a single industry and an autarchic business model (Gittell 1995; Kenny and von Burg 1999; Saxenian 1994). The problem was that the minicomputer industry collapsed much more rapidly than the textile industry a half century before. Furthermore, this time around Lowell was, at least in appearance, virtually a single-company town.[20]

But the story does not end here. Surprisingly to most everyone, both Massachusetts and the Lowell area enjoyed a decade of growth beginning in the early 1990s. The second renewal was not driven by the emergence and rapid growth of big, vertically integrated companies and it was not associated with the rapid rise of a single industry. Out of a population of roughly 120 high-tech companies in the Lowell area in 2003, only one employed over 3,000 and only two others employed over 2,000.[21]

The 60 high-tech companies with 25 or more employees operating in the Lowell area in 2003 are shown in Table 4.1 (another 60 had fewer than 25 employees in 2003). The immediate difference between the first and second high-tech boom decades in Lowell is obvious from a perusal of Table 4.1. Whereas the first renewal was driven by one firm and one sector, the second is better described as "clustered diversity" (Martin 2005: 29), a combination of specialization and diversity. The 60 business units with over 25 employees have been separated into five major technology-defined groups of Instruments and Devices; Optics, Imaging and Photonics; Network Communication Equipment; Software Tools; and Factory Automation Equipment. Collectively these five groups employed nearly twice in 2003 in the Lowell area what Wang Laboratories employed in its peak years of 1985 and 1986.

Where did these companies and associated technology groups come from? In

Table 4.1 Specialist technology groups in the Lowell area

Company	Year* founded	Empl. in 2003	Company	Year founded	Empl. in 2003
Instruments and Equipment					
I. Test and Measurement					
Teradyne/GenRad/ Sierra Research and Technology**	1960	1,200	Thermo-Electron/ KeyTek Instruments	1975	90
ENSR	1968	1,267	Thermo Moisture Systems Corp.	1976	150
Assurance Technology	1969	90	Ionpure Technologies	1989	200
ESA	1970	86	Thermo Detection/ Thermedics Detection Inc.	1991	75
II. Factory Automation Equipment					
Styletec	1970	25	Innovative Products and Equipment	1980	45
Brooks Automation	1978	2,500	Bull Electronics (acquired and merged into Celestica in 2001)	1987	165
III. Microwave Integrated Circuit					
M/A-COM	1958	3,300	Hittite Microwave	1985	150
Microsemi Microwave Products	1985	70			
Optics, Imaging, Photonics					
IV. Optics Tools and Systems					
McPherson	1953	50	Optelic US	1985	70
Minuteman Labs	1967	40	Diamond USA	1990	150
Dielectric Sciences	1970	35	Cynosure	1991	150
Barr Associates	1971	350	Konarka Technologies (photonics)	2001	25
Shafer Corp	1972	44			
V. Digital and Signal Image Processing*					
Bard Electrophysiology	1969	150	Zoll	1980	844
Sky Computer	1980	100	Mercury Computer Systems	1982	600
Network Communications Equipment					
VI. Network Equipment					
Integral Access	1996	70	Astral Communications	1998	180
Nortel Networks	1997	250	SnowShore Networks	2000	45
Netnumber	1997	29	Storigen Systems	2000	65
Captivate Networks	1997	93	Mintera	2000	43
Sonus Networks	1997	497	WaterCove Networks	2000	110
Sycamore Networks	1998	400	Narad Networks	2000	50
Convergent Networks	1998	300	Airvana	2000	100

(Continued.)

Table 4.1 Continued

Company	Year* founded	Empl. in 2003	Company	Year founded	Empl. in 2003
		Software Tool Companies			
VII. Network Communications					
NetScout	1984	335	Quallaby Corp	1996	65
Steleus Inc.	1985	150	Brix Networks	1999	50
Biscom	1986	60	SavaJe Technologies	2000	50
OpenPages	1990	50	Amperion	2001	65
Universal Software	1992	140	Acopia Networks	2002	65
Softlinx	1993	25			
VIII. Enterprise Software Tools					
Kronos	1977	2,375	Datawatch	1985	100
Davox	1981	470	MatrixOne	1994	600
Zuken USA	1983	500	Mission Critical Linux	1999	85
Iris Associates	1984	450			

Notes: *Most companies were founded closer to Boston but subsequently relocated to the Lowell area

** The 1,200 employment number exaggerates the presence of Teradyne and GenRad in the Lowell area. Teradyne acquired Lowell area Sierra Research and Technology (est. 1984, 35 employees) and GenRad (est. 1915, 500 employees in MA) in 2001 and moved its Assembly Test Division from Boston to Westford in 2002 before relocating again to North Reading in 2003

*** All of these companies have medical device products, as do a number of companies in other technology groups including ESA and Teradyne

Source: vTHREAD

fact, many were founded closer to the Boston metropolitan region and moved out after reaching a certain size.[22] But Wang had done the same and the question of technological specialization remains. Why did the high-tech companies which came to locate in the Lowell area have this particular set of specializations? No one has claimed that this pattern or indeed any pattern of "clustered diversity" was policy-driven.[23]

Clearly, the Lowell area is exceptional. But to date the contours of its sectoral composition and the sources of its industrial revival have not been explained. Perhaps the extremes of high-tech growth, collapse, and growth again offer secrets to a better understanding of regional specialization and innovation processes.

We turn now to an exploration of specialization dynamics using the high-tech companies of the Lowell area as an "industrial laboratory." We seek first historic patterns of continuity and change in regionally distinctive production and technology capabilities and second industrial dynamics of differentiation and transformation.

Path continuity dynamics

The concept of path dependence has been widely used by economic geographers to research the evolution of economic landscapes (Martin and Sunley 2006). While the emphasis has been on how regional economies can become locked into paths that lose dynamism, path dependence can also have an upside and contribute to regional industrial renewal. In these cases a region's legacy of distinctive production and technology capabilities are leveraged, integrated, differentiated and reshuffled to create new industrial sectors. In this section we interpret industrial renewal in the Lowell area in terms of specialization dynamics driven by business enterprises and anchored in the region's production and technology capability heritage.

While every successful company has developed a unique concept or product idea to distinguish itself in the marketplace, we can also find common patterns. We look first at common production capabilities and second at common technological capabilities.

Leveraging production capabilities

All 60 companies and every specialized technology group in Table 4.1 can be classified within the broad category of instruments, tools, and equipment. Not a single company does mass production or operates in consumer markets. In this, all of the Lowell area companies reflect New England's specialization in production capabilities that can be traced back to the establishment of the principle of interchangeability at the Springfield Armory in the early decades of the nineteenth century and continuing up to leadership in tools and devices in the life science industries of today (Rosenberg 1972; Best 1990).

The uniqueness, and surprising robustness, of New England's industrial economy has long been obscured by standard classification categories and codes. Since the rise of mass production and the consolidation of the automobile industry in the mid-West, New England's industrial experience has been examined through the lens of mass production. It always comes up short, if not backward. The theme of industrial failure is illustrated with high-volume consumer industries, such as apparel and shoes, which did not survive the development of mass production capabilities in other regions.[24]

The conceptual lens of production capabilities reveals a different industrial landscape. Massachusetts companies specialize in precision instruments, industrial equipment, and complex product systems. In fact, the historic lack of high-volume production capability in New England was countered by the development of distinctive capabilities in complex product systems such as jet engines, missile defense systems, minicomputers, factory automation equipment, and "backbone" switching equipment for the telecommunication carriers.[25] Massachusetts lacks a heritage in mass production engineering, consumer-oriented technologies and associated business organization capabilities; its strengths are in industrial-oriented technologies, precision and systems engineering, complex products and, as will be demonstrated, software tools.

From the production capabilities perspective, the high-tech growth periods did not represent a break with the past. Instead, the successful enterprises leveraged the region's distinctive production capability legacy and, in the process, renewed it. While the production capability threads go back, their applications were often in different and even newly established sectors. In all cases, the success stories have been innovative in new product development and technology management.

This casts a new light on Wang. Officially, the high-tech era in Lowell begins with Wang's move to the city in 1977.[26] Wang, the story goes, represented a break with the "mature" industrial past of the region and signaled the transition to high tech (Glaeser 2005). Of course, this is partly true. But, from the perspective of the region's production capability heritage in instruments, tool making and industrial equipment, the high-tech boom in Massachusetts did not begin as a clean slate or a *tabula rosa*. In fact, Wang's early history involved leveraging and re-"tooling" the State's instrument and tool making heritage. An ex-Wang engineer captures it:

> One aspect of Wang Labs' experience that gave them a "one-up" on their competitors is that the company already had experience with real-time control systems. In the early 1960's, before Wang got into the calculator business and well before the computer business, Wang Labs was involved with Warner-Swasey, a company that made metal working machinery such as lathes and milling machines. Wang developed control systems that would automate the operation of these formerly man-operated machines, allowing faster, more accurate machining of precision metal parts. The work of developing these Numerical Control (or "NC," as the technology came to be known) systems contributed to Wang's later development of control systems based on its calculators.
>
> (Bensene 2001)

The Wang examples illustrates the dynamic specialization theme of continuity and change.[27] Before Wang there were machine tool and precision instrument makers. Wang seized the opportunity to marry machine tools with electronic controls, which was itself an intermediate stage in the marriage of digital computing and control systems and the region's early industrial leadership in cybernetics.[28] Even here, the instrument-making heritage was not lost; it was subsumed in the region's transition to complex product systems, a key element in the rapid growth of the defense industry in Massachusetts (Best 2001: chapter 5).[29] And, as we shall see below, Wang's imprint on the region can be found in companies thriving in the Lowell area today.[30]

Adapting and integrating technology capabilities

The broad category of instruments, tools, and equipment can be broken down into the eight specialist technology groups shown in Table 4.1 by application of the second filter in the technology taxonomy. The Lowell area provides empirical support for the concepts of "related variety" (Boschma and Frenken 2006) or "clustered diversity" (Martin 2005).

Furthermore, while the groups are technologically diverse, they share a common characteristic: they both leverage the region's technology capability heritage and integrate old and new technologies.

Thus instead of being victimized by being locked into computer technology with the collapse of the minicomputer industry, the Lowell area is also home to a small, closely related group that has successfully integrated computer and image-processing technologies. Three (Bard, Zoll, and Mercury Computer Systems) operate in the rapidly growing medical devices sector in Massachusetts and one in detection systems.[31] They represent technology path adaptability in computer technology.

The largest, Mercury Computer Systems, Inc., has 600 employees and was established in 1982. Mercury Computer supplies embedded computer systems for MRI and CAT-scan OEMs (original equipment manufacturers) GE, Philips, and Siemens Medical Systems[32] and for defense applications.[33] It supplied GCA Corporation's wafer stepper in 1984, a machine used to produce precision-aligned sub micron semiconductor components. Mercury Computer Systems has wholly owned subsidiaries in the UK, France, and Japan.

Sky Computer, established in 1980 and employing 100, supplies half the image-signal-processing computers for explosive detection systems that scan luggage at US airports. Sky's software compilers and development tools run on multiple systems in industrial, medical, and defense applications.

The legacy of the computer era is more in software than hardware. Here, too, we find that technology integration involved long-established capabilities. Microwave technology (the radar segment of the electromagnetic spectrum) began in the region during World War Two. It was the core technology that drove the early growth of Raytheon, the largest industrial employer in Massachusetts. Interestingly, the largest and oldest industrial employer founded in the Lowell area is a microwave integrated circuit manufacturer. Founded in 1958 and employing 3,300 in 2003, M/A-COM is the big player in a small group in the Lowell area that contributed to the advance of a technology capability trajectory for half a century. Originally known as Microwave Associates, it provided magnetrons to US Army Signal Corps. Reflecting a move into wireless telecommunications, its name was changed to M/A-COM.[34] More recently, it has enjoyed a resurgence with the growth of mobile telephony.

The region's capability in microwave technology is not lost on leaders in the industry. In 2001, Microsemi Microwave Products, a $200 million public company headquartered in Irvine, CA, established a presence in the Lowell area with the acquisition of New England Semiconductor.[35]

Optics is the specialist technology group with the longest legacy. The development of optics-related capabilities in Massachusetts goes way back to the early days of precision machining and the age of amateur astronomers.[36] Two optics technology companies in Lowell are particularly impressive.

McPherson (est. 1953, 50 employees) has 60 years of optics experience.

McPherson's first product was a spectrometer for Air Force Cambridge Research Lab that was launched into space with an "Aerobe" rocket in 1954. Today it supplies the world's science labs with optics tools or precision measuring instruments. McPherson instruments vary from miniature to versions that weigh over 20 tons and span 70 feet. Its spectrometers measure electromagnetic radiation and enable scientists to investigate small wavelength regions of the electromagnetic spectrum. The Company's spectrographs fly in space rockets and allow scientists to record and search out ancient events in the universe.[37]

Barr Associates, Inc. (est. 1971, 350 employees) designs and manufactures infrared optical filters from less than 200 nanometers' wavelength out to the far infrared (to 35 microns); few companies exist that can meet the challenge of optical filters to these wavelengths. The latest Hubble Servicing Mission involved the replacement of the Faint Objects Camera with a new, faster and more powerful Advanced Camera for Surveys. This new piece of equipment is ten times more sensitive, has a wider field of view and is four times faster at retrieving data. This new instrument contains approximately 25 optical filters designed and manufactured by Barr Associates.

Thus a regional optic and imaging technology capability can be traced back to the early days of Massachusetts' industrial history; microwave technology to the Second World War period; and computer technology to the early postwar period. All have been renewed and adapted and can be found in Lowell area specialist groups of today. The collapse of the minicomputer industry did not signal the end of computer technology capabilities. The basis of similarity of the specialist groups in Table 4.1 varies, and over time specialist groups are reconstituted, emerge, and differentiate.

The regional capability to integrate technologies is central to the success of enterprises in complex product systems.[38] This has been expressed in products as diverse as jet engines, missile defense systems, semiconductor equipment manufacturing, minicomputers, and telecommunication network switching equipment. In all of these cases, the region has demonstrated a distinctive "organizational skills" (production capability) in integrating a range of technologies within a single complex product system. Systems engineering was advanced with every product iteration. While these capabilities and skills were developed in integrated "closed" technology systems, they became a valuable resource for the hundreds of small specialist companies that seized market opportunities created by the transition to open standards and open system business models.

Differentiation and transformation dynamics

The examples in the previous section illustrated path continuity as the leveraging and renewal of a region's distinctive production and technology capability heritage. Here we turn to examples of dynamic specialization which combine capability path continuity with industry differentiation and transformation.

Network communications and software tools are the two largest and fastest growing specialist technology groups in the Lowell area. Both groups illustrate

Jane Jacobs's theme that growth is a qualitative process of differentiation and transformation. Her famous example is the creation of the brassiere as a new product from a fragment of the apparel industry which fostered the emergence of a new industry and eventually a transformation of the New York City apparel industry (1969 51–53).

Jacobs's differentiation theme resonates with Adam Smith's theory of innovation based on increasing specialization.[39] In both cases, previously existing industry activities undergo a process of internal differentiation and specialization. Whereas Smith emphasized the consequences of specialization on process innovation, Jacobs highlights the potential for transformation of an industry based on differentiation and the emergence of new sectors.

The network communication "cluster" is the largest high-tech group in the Lowell area with 25 companies in 2003. It is composed of an equipment and a software tools group. Technological change, vertical disintegration, and deregulation came together in the 1990s to foster the transition of the telecommunications industry from a voice-centric to a data-centric environment. The development of the data-centric Internet was based on packet-switching technology. For data transmission, the legacy circuit networks, which assign a distinct path for the duration of the connection, were inefficient.

In a sentence, the telecommunications retailers (service providers to households and businesses) demanded equipment supply companies that could design and develop complex switches to meet the requirements both of the new platform technology and of compatibility with the installed circuit-switching system over a potentially long transition period. Not every region has had the capabilities to respond.

Telecommunication equipment making was not new to the extended Lowell region. The location in the 1930s of AT&T's 2-million-square-foot manufacturing site in nearby North Andover signaled the establishment of large-scale telecommunication switching and transmission equipment production in the Merrimack Valley.[40] The plant built circuit-switching equipment to optimize massive volumes of voice traffic over the traditional Bell telephone system.[41]

The AT&T plant, which had become a Lucent facility, did not make the technology transition and was shut down at the end of the 1990s. But, like Wang and computer capabilities, the collapse of the Lucent facility did not signal the end of network equipment production capability in the region. Instead the Lowell area became the site of a large group of new, fast-growing companies in the design and development of the Internet infrastructure and the associated transition to the new public network in telecommunications. But instead of one giant switch-making company, a cluster of communication network companies emerged in packet-switching equipment and software systems.

The leader in the establishment of the new network switching equipment industry was Cascade Communications, established in 1990. Cascade designed and developed equipment that would allow installed networks to more efficiently handle data traffic. Its competitive advantage was to offer, in Ed March's words: "a single system platform to support multiple packet communications protocols;

service providers can offer these services through one system without the need to build separate specialized networks" (2003). Cascade's employment grew from 28 in 1993 to 400 by 1996 before being acquired by Ascend Communications of California in 1997.

Equally important, Cascade Communications was an entrepreneurial firm that fostered techno-diversification via spin-offs within the region.[42] A whole "Cascade family of companies" evolved to meet the equipment needs of the new telecommunication service providers created in the wake of deregulation. The range of complementary technology products in which the Cascade "family" of companies operated is shown in Figure 4.1, "Post-Cascade Communications technology differentiation." While most were acquired or otherwise exited the industry, collectively these and associated firms developed a Massachusetts regional capability in IP (Internet Protocol) products and services required to move data, voice, and video over public networks.

The region's technological advantage in network communications equipment both attracted and was reinforced by the entry of foreign telecommunication systems companies such as Seimens, Alcatel, and Nortel as well as Cisco Systems, the Silicon Valley newcomer. All acquired companies along Interstate 495 in Massachusetts.

The equipment makers are not the only enterprises in the region's telecommunications industry. As shown in Table 4.1, the region and the Lowell area are also home to a number of software companies that specialize in network communication systems. Some of these pre-dated the Internet and repositioned into network communications; most are new entrants. Once again we find specialization and diversity.

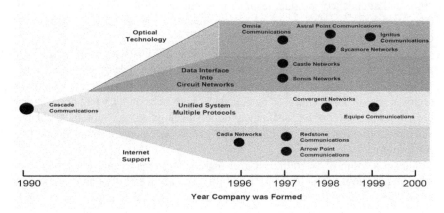

Figure 4.1 Post-Cascade Communications technology differentiation

Source: Edward March 2003

The software tools group is a second example of differentiation and transformation dynamics. It is composed of two sub-groups, network communication software and enterprise software tools. As shown in Table 4.1, both are new industrial sectors constituted primarily by small and medium-sized companies formed in the 1980s and 1990s. All of the 14 network communication equipment companies were founded after 1995. Seven of the software tool companies were established in the 1980s and ten in the 1990s. Several companies grew to employ a hundred to several hundred or more employees including Kronos, Davox, Zuken, NetScout Systems, Universal Software, MatrixOne, Steleus, and Airvana. These companies were growing at exactly the same time that the network communication equipment companies were expanding, and both sectors are software-engineering intensive.

The Lowell area software tool companies can be distinguished from business services in general in that all are high-tech, software-engineering companies. They are functionally analogous to machine tool companies of the region's industrial past. Both specialize in products/services that enhance performance in companies independent of industry.[43] Collectively they supply a diverse range of business services to various industrial sectors. As their performance improves, it reinforces the historic tendency to increased vertical specialization in all business sectors.[44]

Many of the software tool companies began as equipment-making companies. The largest is an example. Kronos, established in 1977, has become a 2,300-employee, $350 million company that began with the founder's idea to put a chip into a mechanical time clock and thereby to bring computer technology to labor and payroll management and production scheduling. Its original attendance-tracking technology platform was an electro-magnetic device; in the late 1980s, Kronos migrated to the PC, and in the early 1990s to LANS (local area networks) and WANS (wide area networks), followed by web-browser, visual factory technology. Along the way customers such as Dell Computer helped to reshape the "time and attendance" product from a management tool for reducing labor-monitoring costs to a workforce empowerment resource consistent with TQM (total quality management) and lean manufacturing.[45]

Most, however, are not large companies. Table 4.2 illustrates an organizational feature of the open system business model: high rates of industrial churn. Churn, in this context, refers to the processes of entry, exit, and growth of firms within a sector or region. High churn, as evidenced by high rates of entry and of enterprise growth is an enabler of regional new product development, technological change, and industrial transformation. Table 4.2 only includes companies with 20 employees or more.

Industrial "churn" is an indicator of regional systems integration capability.[46] Regional economies that have the capability to rapidly reshuffle and reconfigure capabilities and skills to form new teams and take advantage of technology advances or new market opportunities have systems integration and reintegration capabilities.[47]

The principle of systems integration holds that the full effects of an advance in

Table 4.2 Entry and exit of SOF firms in Lowell area, 1996–2000 (employment in firms with over 20 employees)

Company name	Year formed	1996	1997	1998	1999	2000	2001	2002	2003
Voicetek Corporation[48]	1981	120	175	Relocated to CA					
Gartner Group Learning[49]	1987	N/A	66	Relocated to IL					
Mehta Corporation[50]	1991	85	80	33	Relocated to NJ				
Spacetec IMC Corporation[51]	1991	N/A	N/A	76	Relocated to WA				
EDS-Scicon, Inc.[52]	1984	30*	15*	70	100	100	Relocated to TX		
Quickturn Design Systems, Inc.[53]	1993	25	25	39	45	45	45	45	Relocated to CA
Adra Systems, Inc.[54]	1983	190	65	Relocated to other MA location					
ProMetrics Software, Inc.[55]	1982	85	85	10	Relocated to other MA location				
Gulf Computer, Inc.[56]/HCL Tech. (2002)	1979	200	200	200	Relocated to other MA location				
e-StudioLive, Inc.[57]	1971	10*	10*	18*	18	55	52	Relocated to other MA location	
Connolly International, Ltd.[58]	1991	25	25	25	Exit TEL				
Quallaby Corp.[59]	1996	N/A	N/A	N/A	N/A	90	Exit TEL		
NextPoint Networks, Inc.[60]	1996	N/A	N/A	N/A	66	66	Acquired by NetScout		
Iris Associates, Inc.[61]	1984	96	160	350	350	350	500	Acquired by IBM	
ONTOS, Inc.[62]	1985	60	60	60	40	40	Out of business		
Viridien Technologies, Inc.	1997	Entry	N/A	N/A	N/A	120	60	Out of business	
Vikor, Inc.	1999	Entry	Entry	Entry	N/A	N/A	45	14	Out of business
Kronos[63]	1977	N/A	N/A	N/A	N/A	COM	COM	COM	2374
Concerto Software Inc. (PKA: Davox Corp.)[64]	1981	N/A	300	300	300	398	400	470	460
Tek Microsystems[65]	1981	N/A	N/A	N/A	N/A	N/A	N/A	N/A	35
Zuken USA, Inc.	1983	N/A	N/A	N/A	N/A	N/A	500	500	500
NetScout Systems, Inc. (PKA Frontier Software Dev.)	1984	50	140	190	220	220	364	355	340

(*Continued.*)

Table 4.2 Continued

Company name	Year formed	1996	1997	1998	1999	2000	2001	2002	2003
Datawatch Corp.	1985	N/A	N/A	N/A	230	230	175	98	91
Biscom	1986	TEL	TEL	TEL	TEL	TEL	TEL	TEL	60
Digital Voice Systems[66]	1988	N/A	N/A	N/A	N/A	N/A	20	20	20
SoftLinx, Inc.	1993	20	30	30	30	30	40	40	25
Universal Software Corp.	1993	N/A	N/A	N/A	N/A	N/A	140	140	148
MatrixOne, Inc.	1994	N/A	N/A	200	350	366	500	600	450
Trix Systems Inc.	1994	23	12	12	16	16	17	17	27
Steleus Group[67]	1995	N/A	N/A	N/A	N/A	N/A	N/A	N/A	150
Brix Networks	1999				Entry	N/A	N/A	66	65
Mission Critical Linux, Inc.	1999				Entry	75	75	8	8
Airvana	2000					Entry	N/A	N/A	100
SavaJa	2000					Entry	N/A	N/A	50
Amperion	2001						Entry	N/A	65
Acopia	2001						Entry	N/A	65

Notes: Lowell area includes the adjoining townships of Lowell, Chelmsford, and Westford
  N/A: data not available; PKA: previously known as Exit followed by a State abbreviation or company name signifies relocation outside of Lowell area
  TEL and COM signifies the primary industry code for company for the year

Source: vTHREAD

a sub-system are not limited to the sub-system or to the initial and obvious effects; they can include the creation of opportunity to redesign the system as a whole to achieve potentially large architectural change feedback effects.[68] The idea of an "open-systems" business model implies the decentralization and diffusion of design experiments across many enterprises and suggests an extension of the principle of systems integration from a firm to a regional system of mutually adjusting enterprises.

From a regional innovation point of view, the exit of firms is important. Capabilities and skills are released from declining applications and taken up by entering and fast-growing firms. Companies like Digital Equipment Corporation, Wang Laboratories, and Lucent not only advanced technological capabilities, engineering expertise and IT-related skills in the region on a massive scale but created a capabilities and skills resource base that could be tapped by growing firms in succeeding and related sectors. This regional pool of high-tech resources fueled the burst of new firm activity and the human resource requirements of the next wave of high growth companies. It counters the negative lock-in effects of path dependence.

Summary

Regions that specialized in out-dated technologies and declining industries are rarely the location of emerging technologies and new industries. Massachusetts is an exception. Its economic resurgence led by the minicomputer industry was dubbed the "Massachusetts Miracle." In the late 1980s, the State once again lost out to competition from other regions and suffered a collapse of the minicomputer industry. But, just as articles and books were being published on Massachusetts as a case study of industrial decline, a decade-long industrial revival was underway.

The question is why, against the expectations of virtually everyone, did Massachusetts industry enjoy a revival following the collapse of the minicomputer industry?

Industrial renewal and resiliency processes are explored in this paper with the aid of a historical database of high-tech companies of a sub-region in Massachusetts first dominated by textiles and later minicomputers.[69] The two-century industrial history of Lowell is a location in which both industries were established, grew rapidly, and then declined. The crash of Wang Laboratories, located in Lowell, was dramatic and rapid. But, while the City's biggest firm collapsed, the region proved resilient to the new competition of Silicon Valley and elsewhere as new sectors emerged and collectively grew rapidly.

Lowell is the beneficiary of the fecundity of the greater Boston area in the creation of new technologies and enterprises. But the Lowell area has become the address of a series of specialist technology groups which in turn have deepened the region's underlying technological capabilities.

The analysis is consistent with the idea of a regional innovation system as the source of growth in contrast to the technology transfer model of innovation in which local firms commercialize technologies that have been transferred from corporate, university, and government scientific laboratories (Broers 2005; Best 2001). In fact, the transfer metaphor has been appropriate in describing the emergence of the biotech industry. But it does not account for the processes of industrial renewal in the case of the Lowell area. The regional growth process is more complex and interactive.

The first dimension includes the regional production and technology capabilities and skill sets within which firms operate. A key to success was tapping into, leveraging, and extending the region's technological heritage. In fact, the long production capability threads in the form of experiential knowledge, technical expertise, and craft skills is a constant in specialist groups examined; they go back to early-nineteenth-century regional specialization in precision instruments, machine tools, and industrial equipment. We find also technology capability threads in, for example, optics, also going back to the early days but renewed when combined with the emergence of successive new technology knowledge domains in electrical and software engineering. Thus the past has not been lost, although it has been subsumed into more integrative technology knowledge communities.

The organizational characteristics of business and industry are equally central to the renewal and resiliency story in Massachusetts. Here we focused attention on transition from giant companies as the drivers of the growth process to the spread of an "open-system," "focus and network" model of business organization. The spread of the open-systems business model decentralized and diffused the design function and enhanced the new product development capability across a broad band of mutually adjusting enterprises. The example of network communication equipment shows how rapidly new technology mini-clusters can burst onto the scene. And the software "tools" industry illustrates the idea of enterprise "churn," a Schumpeterian process of creative destruction in which resources are reallocated from under-performing to new and fast-growing enterprises.

The new product development process is the mediator between business organization and technological capability. But it is more: it is at the heart of the interactive process by which regional capabilities are deepened and advanced. This is particularly important to an understanding of the cycles of renewal of industry in Massachusetts. Here, at least, the technological capabilities that underlie new product development at the enterprise level and new sub-sector development at the industrial sector level are protean. But their successful development and application depends upon appropriately organized business enterprises. In turn, by designing and developing next-generation product concepts that incorporate technical advances, firms cumulatively and collectively deepen the specialist engineering and craft skill "know-how" in the region. These resources set the stage for future renewals.

What are the implications of the research methodology presented here for the development of a more general theory of industrial renewal and regional resilience?

The firm is the focus of attention because it is the driver of technological change, but not entirely within their own choosing. Instead, they are active agents within a complex industrial system which we characterize in terms of regional production and technological capabilities, and business organization. Together, these two dimensions can under certain circumstances establish a distinct model of industrial innovation: a regional innovation system in which industrial renewal operates at three levels.

At the individual firm level, industrial renewal is a process of successful new product development in response to changing market opportunities. But the new product development process is, at the same time, both a process of upgrading a company's capabilities, including technological capabilities, and a catalyst for cluster churn. These new, upgraded capabilities become resources for the next new product iteration. Successful products become resource magnets for growing firms.

At the inter-firm level, industrial renewal is about systems integration in that elements from established production and technological capabilities are recombined with new or other elements to form new products and sectors. The agency of change is a set of cluster dynamic processes starting with the new product development process of entrepreneurial firms and including enterprise spin-offs as the agent of technology differentiation and increasing specialization, inter-firm networks as the agent of technology combinations, and new firms as the agents of technological speciation, itself enhanced by greater technological diversity. All of these processes are enhanced by the focus and network business model and the open systems model of industrial organization.

Vertical specialization and open-system organizational models are not necessary for industrial renewal, but new company formation is mandatory. The role of new companies here is to force a crisis and response in older companies organized around earlier generation technologies and practices. Ironically, old companies are often haunted by new companies with product ideas that originated within but were not pursued by their own operations at an earlier time. In this way the new companies counter the social and organizational inertia that is inherent in companies that had enjoyed earlier business success. The new companies drag the old to play a role in advancing regional technological capabilities; without the crisis, the older companies can be a barrier to technological innovation and industrial renewal.

At the regional level, the potential for design experiments, capability and skill reconfigurations, and serial regroupings of technology teams is enhanced by an open-system model of industrial organization. But the existence of "open-systems" not only fosters reconfigurations and regroupings, it creates an industrial infrastructure that acts back on capability specialization within and among the constituent enterprises. This specialization, in turn, fosters technological innovation and the potential for yet new enterprises and new configurations of enterprises. In the process, the regional technological capabilities, the source of regional competitive advantage, are revitalized.

Economic downturns, as well, are functional to regional renewal processes. Sustained growth is about capability development with periodic reshuffles forced by economic downturns, market changes, new competitors, or technological change. For this reason the path dependency story needs cluster dynamics, just as cluster dynamics need path dependency of capability development and skill formation.

In all of these ways, the processes of renewal are at the same time the sources of resilience, more so at the regional than the individual enterprise level. The open-system business model is, in effect, an organizational form that turns what would otherwise be enterprise failure to adapt to new competitive pressures into a collective "creative destruction" capability. Resources are shifted from declining to growing products and firms, and the Penrosian firm-learning effect creates opportunities and actions that counter the forces of organizational lock-in that are common to regions with closed-system business models.

Notes

1   Acknowledgements: The author wishes to thank Ron Boschma and Ron Martin for comments on an earlier draft of this paper; Ed March for extensive discussions on the telecommunications network equipment industry as well as comments on earlier drafts; and Albert Paquin and Hao Xie for research assistance.

2   See Cortright (2006) for a review of empirical studies addressing the specialization versus diversity debate.

3   Porter defines clusters briefly as "geographical concentrations of linked industries" and more expansively as follows: "We define a cluster as a geographically proximate group of interconnected companies, suppliers, service providers and associated institutions in a particular field, linked by externalities of various types" (2003: 562).

4   "Regional comparisons offer a huge, almost untapped source of evidence about how our economy really works" (Krugman 1991: 99).

5   When DEC, located in Route 128, for example, designed and developed every major component in the minicomputer, numerous specialist companies emerged for every computer component in Silicon Valley.

6   Japanese consumer electronics companies were leaders in integrating technology management into the new product development process and both into a highly flexible production organization. In brief, they extended the principle of flow from single product to multiple product to new product integration (Best 2001: 40–46).

7   "Clusters are important because of the externalities that connect the constituent industries, such as common technologies, skills, knowledge and purchased inputs" (Porter 2003: 562).

8   Nelson and Winter conceptualized the organization "as a set of interdependent operational and administrative routines, which slowly evolve on the basis of performance feedbacks" and routines as "stable patterns of behavior that characterize organizational reactions to variegated, internal or external stimuli" (Zollo and Winter 2002: 7).

9   The challenges for evolutionary economic geography are to understand the "spatial distribution of routines over time," the "creation and diffusion of new routines in space" and the "mechanisms through which the diffusion of 'fitter' routines occurs" (Boschma and Frenken 2006: 6).

10 In a study of the Netherlands assessing the impact of related variety on growth (Frenken *et al.* 2005), sectors were defined as related variety when they shared the same category in the sector classification scheme at the 5-digit level and as unrelated variety when they belonged to different sector headings at the 2-digit level. The study found that related variety was associated with employment growth in the period 1996–2002.

11 Teece *et al.* (1997) describe dynamic capabilities as "the firm's ability to integrate, build and reconfigure internal and external competencies to address rapidly changing environments." This description substitutes the term, competence for routine but leaves open the questions of what competences are, where they come from, and how they change.

12 For a detailed production capability spectrum that allows for firm and regional comparisons and assessments see Best (2001: Ch. 2).

13 Other examples of regional or national industrial leadership created by the establishment of new principles of production and organization can be found in Best (1990 and 2001).

14 The biological analogy, admittedly imperfect, is to Dawkins's distinction between "replicator" and "vehicle" (1989: 254). The business enterprise, like the body, is a mere vehicle; it is not replicated. Its function is to propagate the replicators. In the natural world, DNA molecules are replicators; they are the fundamental units of natural selection. Distinctive regional capabilities, not the business enterprises, are the basic unit of competitive selection across regions. But, as we have noted, the vehicles may be more or less successful in propagating the replicators.

15 Jacobs distinguishes two growth processes (1969: 129). Preformative growth is a process by which all elements expand by the same increments; epigenetic growth is a process of growth through increasing functional differentiation, much as the human grows from a simple embryo. Economies, she argues, grow by increasing differentiation.

16 Our research method is similar to empirical studies of "revealed" comparative advantage first proposed and conducted by Béla Balassa (1965). In these studies, underlying comparative advantage is interpreted by the examination of traded product statistics. Instead, we use measures of companies and products filtered by a finely granulated technology taxonomy to reveal underlying capabilities which impart competitive advantage to firms in the region.

17 The vTHREAD database is populated with a longitudinal file covering 1989 to 2004, based on CorpTech data. The primary purpose of the CorpTech data set was to provide company information on private and public high-tech companies in the United States. It was supplied quarterly to subscribers and included approximately 55,000 high-tech companies in the United States, including 5,000 in Massachusetts. Although the dataset was not constructed for scholarly purposes, CorpTech established sophisticated data collection and research methodology, including quality control systems and consistency checks. The database is longitudinal: firms in the database were observed and measured over a number of years and their year-to-year records were then linked. CorpTech's data collection methodology had eight phases: company identification using a variety of sources, telephone interview after which the company's products are coded, editing by senior researcher, data entry, internal proofing, machine check which applies numerous tests to each record, external proofing, and written verification of the record from the listed company.

18 Details on the CorpTech taxonomy can be found in Best *et al.* (2004), appendices A and B.

19 Geographers have suggested various "embedding frameworks2 and "learning mechanisms" by which industrial localization and clustering lead to distinct trajectories of innovation. See Bathelt *et al.* (2004). Hopefully application of the capabilities perspective presented here can more specifically characterize the distinctive knowledge domains of regions.

20 "By 1989, approximately one-third (35 percent) of the local labor market's employment was in manufacturing, with industrial machinery accounting for over one-half of the manufacturing jobs. Industrial machinery had become as important to Lowell manufacturing employment as textiles and apparel were at the turn of the century. Further, over 90 percent of the employment in industrial machinery was in one industry, office and computing equipment (SIC 357), which includes minicomputers. One firm, Wang Labs, accounted for the bulk of the local jobs in that industry" (Gittell 1995).

21 We define the Lowell area as the three adjoining townships of Lowell, Chelmsford, and Westford.

22 Lowell has benefited from its location at the intersection of Interstate 495, an outer ring road around Boston, and Highway 3, a short drive to New Hampshire.

23 Many of the companies have been around awhile. Of all 120 high-tech companies in the Lowell area in 2003, 32 were founded in the 1990s. Thirty-eight were founded in 1980s, but 23 go back to the 1970s, eight to the 1960s and five to the 1950s. Of the 60 with 25 employees or more, ten were established in the 2000s, 18 in the 1990s, 16 in the 1980s, nine in the 1970s, five in the 1960s, and two in the 1950s.

24 The claim that mass production capability is extremely rare in Massachusetts is resisted by many in Massachusetts, even many with technology backgrounds. They cite plants such as Gillette, Norton Abrasives (now a division of Compagnie De Saint Gobain of France), and Smith&Wesson. Unless they have been transformed recently, these are not examples of mass production but of mass batch production; these two production systems operate according to different principles (Best 1990: 147–161, 2001: 28–40). While many Massachusetts manufacturing plants have made impressive advances toward "world class manufacturing" performance standards in cost, quality, and time, precious few have been transformed from "mass batch" production methods to multi-product flow, cellular layout, and the synchronization of cycle

times. Each company may have good commercial reasons to stick with mass batch production methods; in addition, they will not easily find the 'deep craft skills' in the region for, or even an understanding of, mass production methods.

25   For example, and as will be shown below, Massachusetts companies supply complex switching gear, infrastructure equipment to the major telecommunication carriers, whereas Silicon Valley companies dominate the high-volume, consumer electronic end of the market such as hubs and routers for business and homes (Best *et al.* 2004).

26   "Wang's greatest gift to Lowell was its address," states William Taupier, the Lowell city manager who facilitated the move.

27   Wang never sought venture capital but, facing a cash crunch in 1959, it sold stock worth 25 percent of Wang Laboratories to Warner & Swasey in exchange for $50,000 and access to $100,000 in loans (Rosegrant and Lampe 1992: 116).

28   For an excellent technical and historical account of control systems, see Mindell (2002).

29   Seen from the perspective of a heritage in industrial equipment, the computer revolution in Massachusetts was not a radical break with the past. In an important way it can be read as a renewal of the region's technological capabilities.

30   The Wang effect has long outlived the company's existence in Lowell. Wang Laboratories not only directly trained thousands in hardware and software technological processes who have gone on to other firms but, according to Chancellor William Hogan, An Wang played a pivotal role in curriculum development at the University of Massachusetts Lowell by pressing for a major expansion in software engineering in the early stages of the minicomputer industry.

31   See Best (2006) for a case study of the rapid growth of the Massachusetts medical devices industry.

32   Others in this category include Bard Electrophysiology (est. 1969, 150 employees) which designs and makes specialist computer hardware geared to electrophysiology for the medical industry.

33   Founder James R. Bertelli was with Analogic, a leader in similar imaging processing technologies. Data General provided start-up funds for Mercury. Mercury evolved from producing board-level hardware to providing a total system solution: "product includes a combination of hardware and software ... by some of the most skilled signal and image processing applications and systems engineers in the industry." The Company's website further refers to: "an unmatched degree of expertise in technologies such as radar and sonar signal processing, digital X-ray and computed tomography (CT), and audio and video image compression, decompression, and reconstruction."

34   M/A-COM is an example of a company that is hard to slot into a single primary technology code. Over time the Company has progressed from being primarily a defense contractor, to a component producer, to telecommunications. It has always been a microwave integrated circuit designer and developer.

35   Microsemi also acquired Compensated Devices, Inc. of Massachusetts in August 2001 and, along with Microsemi's Watertown, Massachusetts operation, consolidated both in a newly outfitted facility in Lowell. www.microsemi.com/finance/presentations/AnnualReport2002.pdf.

36   Other interesting and still operating optics companies in terms of history or attraction of foreign investment include American Optical Lens Company (est. 1832, 40 employees), located in Southbridge, MA; The O.C. White Co., a manufacturer of microscopes (est. 1894, 10–24 employees), located in Three Rivers, MA; and Dolan-Jenner Industries, Inc. (est. 1962, 120 in nearby Lawrence, MA). Dolan-Jenner makes optical inspection equipment, gauging systems, and fiber optic illumination systems for multiple industries. Its parent is the Danaher group, which includes Kollmorgen Electrical Optical. More on the early links between skill machinists and the early optical industry in Massachusetts can be found in Best (2001: 136). Foreign investment

in photonics is considerable in Massachusetts. Of 55 photonics CorpTech-listed companies in Massachusetts with sales of over $5 million, 14 are operating units with headquarters in foreign countries: Germany 5, Japan, Holland 2, Italy, UK 2, Canada, France, and Switzerland.

37 A unit of McPherson, Inc., Space Optics Research Labs. (est. 1962, 20 employees), manufactures optical components and systems including laser and optical equipment, and supplies optical manufacturing services to government and aerospace.

38 The greater Massachusetts region is unmatched in 'technological' systems integration capabilities that cut across computers, communications, and control systems, a heritage that predates 'organizational' systems integration. The region's specialization in 'systems integration' is traceable in part to the demand for systems engineering generated by the Electronics Systems Center (ESC) at nearby Hanscom Air Force Base in Lexington, Massachusetts. The ESC is the US Air Force's site for C4I, a defense industry acronym that stands for command, control, communications, computers, and intelligence. The ESC has managed nearly 200 C4I projects going back to the 1950s and the Semiautomatic Ground Environment (SAGE) project, an air defense system involving the integration of technologies underlying missiles, radar networks, gunfire control, guidance systems, and high-speed digital computers (Hughes 1998: 17). The ESC followed SAGE with management of the Airborne Warning and Control System (AWACS) in the 1970s. Thus the ESC has been a project manager for an integration of the region's historic capability in complex product systems, such as jet engines and telecommunication switching equipment, with software systems control.

39 Smith linked specialization to the discovery process (Loasby 1991).

40 This is a study in industrial renewal on its own as AT&T's location decision was based on the availability of a large industrially experienced workforce resulting from the decline of the footwear industry in the region (Ed March 2003—see Note 41).

41 This section draws heavily from an Anticipating Technology Trends internal document written by Ed March and titled "Cascade Communications 'Family of Companies,'" August 2003.

42 Details on each of the 11 Cascade "family of companies" can be found in March 2003.

43 Like the region's machine tool companies of the past, many of the software tool companies operate in global markets. A Japanese company, Zuken, established a Lowell Area unit Zuken USA in 1983 which grew to 500 employees. It specializes in concurrent design tools, including CAD/CAM (computer-aided design/computer-aided manufacture) software for multiple brand workstations, and sells to the engineering industry. Zuken USA has PCB (printed circuit board) design service bureaux in 10 countries and hundreds of 'partners." Established in 1994, MatrixOne, Inc. grew from 200 to 600 employees between 1999 and 2003 by offering product lifecycle management (PLM) services to manage global supply chains. It partners with Tata Consultancy Services (Asia's largest software company) to establish a PLM Center of Excellence to offer *Fortune* 500 companies worldwide assistance in reducing new product development cycle times.

44 In the early days of the industrial revolution, specialist machine tool-making companies spun out of vertically integrated companies to form new groups of specialist machine tool companies. Analogously the information technology revolution has been accompanied by increasing vertical specialization accompanied by the emergence of software tool companies.

45 Two other examples are Davox and Datawatch Corporation. Davox, established in 1981, (recently renamed Concerto Software, Inc.) began as a supplier of data concentration equipment and computer-voice telephone equipment primarily to call centers. With the acquisition of AnswerSoft, Davox enhanced customer interaction features and entered the Customer Relations Management (CRM) market with a product that combines telephony, fax, email, and the web. It employs nearly 500 with 300 in

Massachusetts and sells to financial institutions, airlines, telecommunication providers, insurance companies, utilities, and retailers, and has subsidiaries in Canada, the UK, Mexico, Germany, Brazil, Japan, and Singapore. Datawatch Corporation began in 1985 as a designer and manufacturer of computer workstations for high security applications but pursued opportunities in business software by developing a distinctive capability in the use of information technologies for report writing. Rather than pursing information storage technologies, a Massachusetts specialty, Datawatch chose to focus on information use. This includes data extraction, mining and transformation across disparate databases and multiple platforms.

46 This churn of enterprises counteracts the 'innovator's dilemma' of single companies as described by Clayton Christensen (1997), but only if the region is populated by the 'open-system' or focus and network business model. Regional technology capabilities are secure if the regional system of enterprises includes both incumbents and attackers. In contrast, a region in which free entry is limited risks blocking the entry of firms with disruptive technologies much like the Upas tree poisons the seedlings of other species of plants around it; for the example of heavy engineering killing alternative technologies and regional growth in Glasgow, Scotland, see Checkland (1981).

47 Churn is a regional organizational response to the inherent limits of even brilliant innovators to predict the technological future combined with the inherent uncertainly of the specific technological future. The challenge is captured by Paul Severino, a serial entrepreneur in Boston's Route 128 region: "Ken Olsen [founder of DEC] was and is brilliant but one man can not always guess right about the future."

48 Acquired and merged into Aspect Telecommunications Corp., San Jose, CA in 1998.

49 Moved to Lowell area in 1997; acquired and merged into Illinois-based National Education Training Group in 1998.

50 Moved to NJ in 1999.

51 Acquired and merged into Labtec Enterprises Inc., Vancouver, WA in 1999.

52 Moved to Lowell area in 1998; acquired and merged into EDS Corp. in 2001.

53 Acquired and merged into Cadence Design Systems, CA in 2003.

54 Moved to Tewksbury, MA in 1998.

55 PKA Digitech; acquired by parent ProMetrics Ltd. in 1991. UK Moved to Athol, MA in 1999; 10 employees from 1999 to 2003.

56 Parent is HCL Technologies, India. Both LangBox Division (founded 1994) and Spartacus Technologies Division (founded 1981) moved to Quincy, MA, in 1999. Renamed HCL Technologies (Mass.) in 2002.

57 Moved to Lowell area from Burlington, MA in 1999 and relocated to Tewksbury in 2003 and downsized to 30 employees.

58 Twenty-five employees every year between 1996 and 2002, and 14 in 2003; industry code changed from Software to Telecommunications in 1998.

59 Moved to Lowell area in 2000 from Burlington, MA; became TEL in 2001.

60 Acquired and merged into NetScout in 2001.

61 Merged into IBM in 2002.

62 Moved to Andover, MA in 2001 and out of business in 2002.

63 Moved to Lowell area in 2000 from Waltham, MA and repositioned from COM to SOF in 2003. Employed 1,200 in 1996 and 2,000 in 2001.

64 Moved to Lowell area from Billerica, MA in 1993.

65 Moved from Burlington, MA in 2003.

66 Moved to Lowell area from Burlington, MA in 2001.

67 Acquired by Tekelec of CA in 2004.

68 Henry Ford, for example, did not see the introduction of electricity as a reduced cost of energy; he saw it as an opportunity to create a new production architecture. Ford's engineers redesigned the plant layout from functionally specialized departments to continuous flow lines. This demanded reengineering not only of the production process but the machines and work activities around a new principle (Best 1990).

69 Unfortunately, the company dataset that populated vTHREAD was discontinued in 2005. Thus it is not possible to examine the effects of the Great Recession beginning in 2007 on Lowell industrial activity. We can say that there is no a priori evidence that the industrial innovation and sectoral transition dynamics examined here have declined in force.

Bibliography

Baden-Fuller, C. and Winter, S. 2006. "Replicating organizational knowledge: principles or templates," DRUID Summer Conference 2006.

Balassa, B. 1965, "Trade liberalization and revealed comparative advantage," *Manchester Journal of Economics and Social Studies*, 33 (2): 99–123.

Bathelt, H., Malmberg, A., and Maskell, P. 2004. "Clusters and knowledge: local buzz, global pipelines and the process of knowledge creation," *Progress in Human Geography*, 28: 31–56.

Bensene, R. 2001. "Wang Laboratories: from custom systems to computers," October, updated June 2002, www.oldcalculatormuseum.com/d-wangcustom.htm.

Best, M. 1990. *The New Competition*, Cambridge, MA: Harvard University Press.

Best, M. 2001. *The New Competitive Advantage: The Renewal of American Industry*, Oxford: Oxford University Press.

Best, M. 2006. "Massachusetts medical devices: leveraging the region's capabilities," *Mass*Benchmarks, 8 (1): 14–25. Online at www.massbenchmarks.org/publications/massbenchmarks.htm.

Best, M., Paquin, A., and Xie, H. 2004. "Discovering regional competitive advantage: Massachusetts high-tech," Online publication of the Business History Association, www.thebhc.org/publications/BEHonline/2004/BestPaquinXie.pdf.

Boschma, R.A. 2004. "The competitiveness of regions from an evolutionary perspective," *Reg. Studies*, 38 (9): 1001–1014.

Boschma, R.A. and Frenken, K. 2006. "Why is economic geography not an evolutionary science? Towards an evolutionary economic geography," *Journal of Economic Geography*, 6: 273–302.

Broers, A. 2005. *The Triumph of Technology: The BBC Reith Lectures 2005*, Cambridge: Cambridge University Press.

Checkland, S.G. 1981. *The Upas Tree: Glasgow 1875–1975*, Glasgow: Glasgow University Press.

Christensen, C. 1997. *The Innovator's Dilemma: When New Technologies Cause Great Firms to Fail*, Boston, MA: Harvard Business School Press.

Cortright, J. 2006. *Making Sense of Clusters: Regional Competitiveness and Economic Development*, Washington, DC: Brookings Institution.

Dawkins, R. 1989. *The Selfish Gene*, 2nd edition, Oxford: Oxford University Press.

Flynn, P. 1988. Lowell: "A High Tech Success Story," in Lampe, D., *The Massachusetts Miracle*, Cambridge, MA: MIT, pp. 275–294.

Frenken, K., Van Oort, F.G., Verburg, T., and Boschma, R.A. 2005. "Variety and regional economic growth in the Netherlands," *Papers in Evolutionary Economic Geography*, 05.02, Utrecht: Utrecht University.

Gittell, R. 1995. "The Lowell high-tech success story: What went wrong?—Lowell, Massachusetts," *New England Economic Review*, March–April.

Glaeser, E. 2005. "Reinventing Boston," *Journal of Economic Geography*, 5 (2), pp. 119–153.

Glaeser, E.L., Kallai, H.D., Scheinkman, J.A., and Schleifer, A. 1992. "Growth in cities," *J. Pol. Econ.*, 100 (December), 1226–1252.

Grove, A. 1996 *Only the Paranoid Survive*, New York: Doubleday.

Henderson, V. 1997. "Externalities and industrial development," *Journal of Urban Economics*, 42: 449–479.

Henderson, V. 2003. "Marshall's scale economies," *Journal of Urban Economics*, 53 (1): 1–28.

Hughes, T. 1998. *Rescuing Prometheus*, New York: Pantheon.

Jacobs, J. 1969. *The Economy of Cities*, New York: Random House.

Kenny, M. and von Burg, U. 1999. "Technology, entrepreneurship and path dependence: industrial clustering in Silicon Valley and Route 128," *Industrial and Corporate Change*, 8, (1): 87–88.

Krugman, P. 1991. *Geography and Trade*, Cambridge, MA: MIT Press.

Loasby, B. 1991. *Equilibrium and Evolution: An Exploration of Connecting Principles in Economics*, Manchester: Manchester University Press.

Loasby, B. 1999. *Knowledge, Institutions and Evolution in Economics*, London: Routledge.

March, E. 2003. "Cascade Communications 'Family of Companies'", Anticipating Technology Trends Research Paper, August.

Martin, R. 2005. "Thinking about regional competitiveness: critical issues," working paper.

Martin, R. 2006. "Path dependence and the economic landscape," in Berndt, C. and Glückler, J. (eds), *Denkanstöße zu einer anderen Geographie der Ökonomie* (Reflections on Heterodox Economic Geography), Bielefeld: Verlag.

Martin, R. and Sunley, P. 2006. "Path dependence and regional economic evolution," *Papers in Evolutionary Economic Geography*, 06.06, Utrecht: Utrecht University.

Mindell, D. 2002. *Between Human and Machine: Feedback, Control, and Computing before Cybernetics*, Baltimore, MD: Johns Hopkins University Press.

Nelson, R. and Winter, S. 1982. *An Evolutionary Theory of Economic Change*, Cambridge, MA and London: The Belknap Press.

O'Connell, M. 1991a. "CEO says Wang can get back in black," *The Lowell Sun*, April 25.

O'Connell, M. 1991b. "Miller: more Wang cuts ahead," *The Lowell Sun*, July 19.

Penrose, E. 1995 [1959]. *The Theory of the Growth of the Firm*, 3rd edn, Oxford: Oxford University Press.

Porter, M. 1990. *The Competitive Advantage of Nations*, New York: Macmillan Press.

Porter M.E. 2003. "The economic performance of regions," *Reg. Studies*, 37 (6&7): 549–578.

Prencipe, A. 2000. "Breadth and depth of technological capabilities in complex product systems: the case of the aircraft engine control system," *Research Policy*, 29: 895–911.

Richardson, G. 1972. "The organization of industry," *Economic Journal*, 82, September.

Rosegrant, S. and Lampe, D. 1992 *Route 128: Lessons from Boston's High-Tech Community*, New York: Basic Books.

Rosenberg, N. 1972. *Technology and American Economic Growth*, New York: Harper and Row.

Saxenian, A. 1994. *Regional Advantage: Culture and Competition in Silicon Valley*, Cambridge, MA: Harvard University Press.

Teece, D., Pisano, G., and Shuen, A. 1997. "Dynamic capabilities and strategic management," *Strategic Management Journal*, 18 (7): 509–533.

Xie, Hoa 2003. *New Firm Creation in Telecommunications and Internet Sector:*

Massachusetts 1996–2002, Master's thesis, Regional Economic and Social Development Department, University of Massachusetts Lowell.

Zollo, M. and Winter, S. 2002. "Deliberate learning and the evolution of dynamic capabilities," *Organization Science*, 15 (3): 339–351.

5 From recession to re-industrialization

A case study of employment changes in North Carolina

Ian Taplin and Minh-Trang Thi Nguyen

Introduction

Since the 2007/2008 financial crisis there has been considerable discussion in the press regarding the viability of an apparent resurgence of manufacturing in the USA, even in industries such as textiles that had traditionally been viewed as 'sunset' industries (McWhirter and McMahon, 2013). However, whilst there are examples of companies that have seen productivity increases, they have not necessarily been accompanied by wage (Lowrey, 2013) or employment growth (Rattner, 2014), This has led to others questioning whether the apparent manufacturing revival is enough to restore the US economy (Rattner, 2014), or if the business climate can be made more conducive to long-term competitiveness (Porter and Rivkin, 2012). Furthermore, recent data on wage trends suggest that restored US competitiveness is often sustained by low entry-level hourly wage rates: for example $14.50 for auto assembly line workers, with rates of $27 an hour in VW's new Chattanooga plant compared to an average of $67 an hour in Germany (Rattner, 2014).

A recent report by the Boston Consulting Group (Sirkin *et al.*, 2011) argues that a return of manufacturing to the USA is possible precisely because of yearly wage and benefit increases of 15 to 20 per cent in China and higher transportation costs and supply chain risks in Asia. Meanwhile, certain southern states in the USA, with low tax rates, union-free workforces, good infrastructure and low entry-level wages when combined with generous subsidies from state and local government, make the area an attractive proposition for new manufacturing. Does this mean that the path towards re-industrialization in the USA is one built upon low-waged labour – another chapter in the proverbial 'race to the bottom'? Or is it conceivable that a new form of manufacturing is becoming the norm, one based upon skilled, more technically mediated work, often found in smaller enterprises that receive less publicity about their operations?

In order to attempt an answer to these questions, this chapter examines structural changes in North Carolina (NC) over the past few decades, paying particular attention to the recent revival of key manufacturing areas. In the context of increased globalization of production, we examine job decline and the recent reversals in that decline in the traditional sectors of textiles and furniture manufacturing. We also comment on the secular growth of non-manufacturing (notably the service sector and specialist agriculture). We look at what this means for types of jobs that are being created and their wage rates. In the context of the debate on skill mismatches, we look at the role of institutional actors (the state, via technical colleges) as agencies for skill creation and the extent to which such policies in a neo-liberal, free market economy will often be viewed with caution. Relatedly we consider attempts by key firms to collaborate with the state to secure appropriately trained workers, but also ponder the extent to which firms are attempting to outsource training in order to gain cost savings in the manufacturing process. We also note the financial support provided by states to attract new businesses and how these subsidies are an increasingly important part of firm location strategies.

Context and issues

With the recession that followed the financial crisis of 2007, there has been much hand-wringing over high rates of unemployment and the difficulties that young workers face in finding full-time employment. The mortgage crisis depressed house prices in many areas, with current estimates of one in five American homes still valued at less than the mortgage held on the property. This situation curtailed geographic mobility, preventing workers from relocating to areas where jobs might be available. Median family wages have changed little in the past decade as globalization placed American workers in competition with those in low-wage emerging economies. Consumer spending has been flat for the last five years, and even the semi-annual orgy of retail therapy that accompanies 'Black Friday' (the day after Thanksgiving) and December 26th has failed to alleviate retailer concerns about their diminished margins. Concern over stymied economic growth bodes ominously for individual workers and politicians, both of whom bear the brunt of the crisis, albeit in different but related ways. And yet, at the same time, many businesses have prospered, profit margins are up and the days of executive bonuses are returning to their pre-2007 levels. How might one explain this apparent paradox of corporate success and individual worker despair?

One answer can be found by looking at the action of one famous American company, Caterpillar Inc., in January 2012. Following a strike and lock-out at one of their plants in Ontario, Canada, where workers had rejected the lower wages that were on offer with a new contract, the company argued that, without such a concession, they would be forced to relocate to an area where wage rates were cheaper and labour more compliant. Many initially thought that they were proposing to move to Mexico or China. In fact they were citing wage and benefit costs at their rail-equipment plant in LaGrange, Illinois, which were less than half that of the company's locomotive assembly plant in London, Ontario (Hagerty and Linebaugh, 2012). Japan's Bridgestone Corp. recently made a $1.1 billion investment to expand and build a new car tyre plant in South Carolina; similarly Boeing Corp. recently opened their new jetliner assembly plant in

Charleston, SC. In both cases, lower wage rates, non-unionized workforces, flexible work practices and high rates of worker productivity have combined to enhance company competitiveness.

Another aspect of the revived competitiveness of US manufacturing lies in capital investments and automation. Whirlpool Corp. just announced that it is moving some of its washing machine production from Monterrey, Mexico to an existing plant in Clyde, Ohio, creating 80–100 jobs at the latter site (Hagerty, 2013). Even though wages at the Ohio plant ($18–$19 an hour) are approximately five times higher than in Monterrey, the company claims that a combination of automated production, much lower electricity costs, transportation cost savings and improved logistical coordination will lower their overall costs. As part of efforts to bring actual manufacturing closer to markets where products are sold, firms such as Whirlpool are trying to be more responsive to changes in demand.

As of 2010 US manufacturing labour costs per unit of output were 13 per cent lower than a decade earlier, whereas in Germany costs have risen by 2.3 per cent and in Canada by 18 per cent. When one adds together lower energy costs (the result of shale gas production increases), more productive workers, plus a weaker dollar, US manufacturing has become increasingly viable despite its location in a nominally high-wage country. Not surprisingly, some jobs that had been outsourced to Asia in the 1990s are now returning to the USA, taking advantage of enhanced workforce flexibility as well as avoiding logistical problems in supply chain management. Wage and benefit increases of 15–20 per cent a year in China's factories, without concomitant productivity increases, continue to reduce that country's labour cost advantage, especially for low-value-added manufacturing. When one then adds transportation costs, import duties, energy costs and real estate inflation, the cost savings for manufacturing in China are minimal (Sirkin *et al.*, 2011).

While some manufacturing jobs continue to move to offshore plants, since 2010 companies have created 80,000 manufacturing jobs in the USA by moving production from foreign countries (Hagerty, 2013). Furthermore, companies are capitalizing upon additional technological innovations which allow them to keep design and initial production work in-house, cutting out suppliers and discouraging subsequent manufacturing that might have been done overseas. As a consequence, cross-border investment outflows fell by 18 per cent in 2012, with similar drops anticipated by the end of 2013.

The issues delineated above – productivity enhancing technological innovations, supply chain rationalization, comparatively low wage rates for US workers, lower overhead costs and finally, changing patterns of trade – are part of a broader picture that helps us understand the changing nature of the American workplace. Since the 1980s, much has been written about de-industrialization in the US and other advanced industrial societies (Bluestone and Harrison, 1982; Cohen and Zysman, 1987; Dicken, 2007), as manufacturing jobs moved offshore when (mainly Asian) emerging economies embraced an export-led manufacturing model of economic growth. When combined with the strengthening of the dollar during the 1980s, growth of the internet, worker demographics (and work preferences) plus the general globalization, the result was a significant decline in manufacturing employment, from 22 per cent of all non-farm payrolls in 1977 to 9 per cent in 2012. (US Bureau of Labor Statistics, 2013).

The partial reversal of this trend since 2012, however, is indicative of a possible re-industrialization, albeit predicated on a revitalized manufacturing sector in which higher-value-added work is performed by highly skilled workers. In the USA, capital investments to improve worker productivity in the years after 2008 are an important component of this trend; for example capital expenditures have increased by 31 per cent since 2009 (Waters, 2011). But so is a decline in the value of the dollar and falling unit labour costs – partly a function of subdued wage rates of the type alluded to in the Caterpillar example. Firm output growth is no longer solely productivity based; it is now complemented by new job creation. Manufacturers are adding workers, but what is often distinctive about these new jobs is the requisite level of skill. Firms are seeking workers with numerical and techno-engineering skills that allow them to operate computerized machinery; to be able to troubleshoot production problems; and to be able to multi-task in different aspects of the production process. In other words, they require post-secondary education with a solid grounding in basic engineering and computer/design capabilities, but not necessarily a college degree.

This has raised questions about the availability of such labour as firms are increasingly citing the lack of skilled workers for their limited hiring. More and more firms are claiming that they are unable to hire workers with appropriate skill sets for the type of skilled manufacturing jobs they are creating (North Carolina Commission on Workforce Development, 2007, 2011). Yet it is difficult to disentangle apparent employer concerns over skill shortages and their desire to hire more workers with comments and actions that possibly belie wage-depressing strategies. Given income trends in the USA, it would appear that the possession of skills does not always translate into high wages, at least not for manufacturing workers. Either that, or firms are looking, often unsuccessfully, for skill sets that are in short supply and hence are not able to hire workers in sufficient numbers to have an impact upon aggregate wage trends.

North Carolina

Up until the 1980s, North Carolina had a strong manufacturing base in textiles/apparel and furniture, together with a vibrant rural sector based upon tobacco processing and chicken/pig farming. Twenty per cent of state income came from manufacturing compared to a national average of 12 per cent. See Figure 5.1 for details of sectoral employment breakdown for NC. The state was heavily dependent upon this sector, so any decline or job loss therein would have dramatic consequences for employment rates. Dispersed regional centres for health care and banking provided high-paying employment for skilled professional workers. Otherwise, income levels were lower than national averages. Also much of the manufacturing was based in small towns where a company would be the principal employer and hence arbiter of local economic activity.

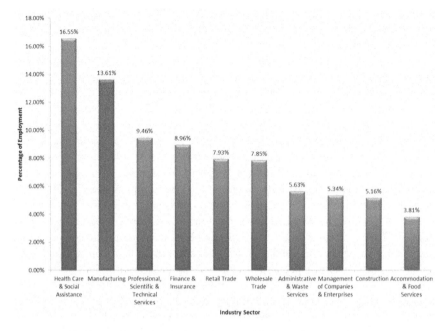

Figure 5.1 North Carolina – employment by top 10 industrial sectors, 2011

Source: US Bureau of Economic Analysis, 2011

During the 1980s, with the exception of some rural counties, unemployment rates of 2–3 per cent prevailed. These were the boom years, with growth in both traditional manufacturing sectors as well as new areas such as health care and banking, finance and insurance. Many textile and apparel mills had 'help wanted' signs posted, often for semi-skilled jobs as sewing machine operators. The rise of the service sector in urban areas meant that more and more semi-skilled jobs in health care, insurance, banking and retail became available. These jobs were appealing to young females who hitherto might have found employment in traditional manufacturing sectors, but found 'new economy' jobs more appealing and less stigmatized – even though they probably paid the same hourly wage rates (Taplin, 1994). The fact that semi-skilled manufacturing wages were similar to those in the emerging service sector where wage rates are traditionally low, and that manufacturers appeared unwilling (or unable) to pay higher wages, is indicative of the extent to which the latter clung to a low-wage strategy.

The declining pool of pliant, young female workers who now had alternative employment opportunities came at the same time as the emergence of first, Central American and then, Asian economies, with economic growth policies predicated upon traditional sectors such as textiles and apparel. The significant growth of imports from these regions followed increased liberalization of trade and eventually China's entry into the World Trade Organization (WTO) in 2001.

These countries possessed abundant supplies of low-skilled and low-waged workers who were perfectly suited for labour-intensive manufacturing of the type done in southern states such as NC. This gradually eroded the competitive position of many domestic US manufacturers and put further pressure on their operating costs, particularly wages and benefits. By the 1990s, some firms in these sectors attempted to further rationalize and automate production, gain productivity increases through work restructuring or concentrate on higher-value-added products (Taplin, 1995). But, for many, the solution was to shut down domestic operations and outsource production to low-wage overseas sites. This proved to be a significant loss for low-skilled workers in areas such manufacturing, construction, and distribution who proved to be the ones most likely to need retraining to find work (North Carolina Commission on Workforce Development, 2011). Some of the biggest losses however, occurred in the furniture and textile/apparel industries.

Between 1996 and 2006 NC textile and apparel employment declined by 65 per cent, from 233,715 workers to 80,232, and the number of plants decreased by 40 per cent from 2153 to 1282. The decline was particular rapid in the period 2002–2005, when textiles employment shrank by 30.8 per cent and that in apparel by 26 per cent (www.soc.duke.edu/NC_GlobalEconomy/index.shtml). A similar story played out in the furniture industry where, during that same short period, 12.4 per cent of jobs were lost (Nwagbara *et al.*, 2002; US Bureau of Economic Analysis, 2007). In all cases, firms went in search of lower production costs that could be found increasingly in China, where hourly wage rates of $0.69 were a dramatic contrast to $14.24 in NC (www.soc.duke.edu/NC_GlobalEconomy/index.shtml). Even with lower productivity rates and higher transportation costs, China proved a very attractive option for industries that relied upon semi-skilled, labour-intensive production processes.

The repercussions of the 2007 financial crisis further exacerbated an already downward trend in manufacturing employment. By 2008 unemployment rates in most areas of the state were around 10 per cent, with rural counties that were heavily dependent upon one or two major factories/plants particularly vulnerable. Almost every month came announcements of layoffs at furniture or textile manufacturers in small towns scattered throughout the state. This pattern was consistent with many areas of the country, with states facing fiscal crises due to declining tax revenues that led to further job losses, particularly for many public sector workers.

While unemployment rates continue to be high (8–10 per cent for most of the state), in recent years many businesses have begun to prosper, including those in the manufacturing sector. The state's manufacturing output fell slightly after the 2007 record of $72.9 billion, but has grown since 2008 to a further record of $88.2 billion in 2012, despite losses in textiles and furniture. At the same time, nonmanufacturing sectors (finance and insurance, health care, real estate, rental and leasing) have grown at a faster rate but have less of a job-effect than manufacturing jobs. Manufacturing is the second largest employment sector in NC, employing 13.61 per cent of the state's workforce, yet leads all industry sectors in its contribution of 19.35 per cent to state gross domestic product (GDP). Figure 5.2 shows the relationship between aggregate GDP growth and manufacturing's GDP.

This growth, however, has not always been accompanied by job creation; initially it appeared to be the result of productivity increases following capital investments (North Carolina Commission on Workforce Development, 2011). In other words, firms found ways to better utilize their existing workers, improving their efficiency and even effectiveness (Taplin, 2006). It is also a function of rising wage rates in China, continued wage inflation in Asia and higher transportation costs – all of which make outsourcing production to this region more expensive. Meanwhile, production costs in the USA continue to benefit from lower utility costs, more efficient supply chain management and technological innovations that enable firms to better coordinate supply with demand. A lower value dollar results in comparative advantage for manufacturers who export; and wage stagnation means unit labour costs have not risen in comparison to other high-wage economies. Manufacturing sector wages in the USA are now only 20 per cent higher than in China and the latter has seen 15–18 per cent wage inflation for the past five years (Sirkin *et al*, 2011). The more that Chinese firms attracted new investment, the greater the demand for employees, but the labour market pool has been shrinking, making workers more expensive to hire.

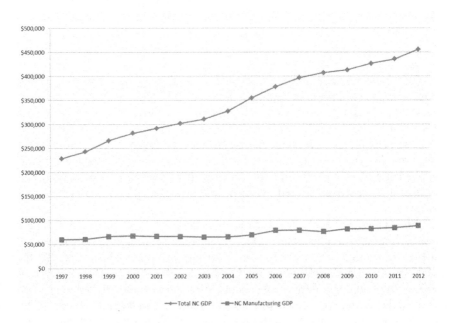

Figure 5.2 North Carolina total GDP and manufacturing GDP ($m)

Source: US Bureau of Labor Statistics, 2013

Although some industries have shifted production to other low-cost Asian sites such as Bangladesh, which is now second to China as an exporter of garments (Al-Mahmoud *et al*. 2013; Davidson, 2013), others have rediscovered the benefits of US production. In recent years, hiring has picked up as firms appear willing to add workers to fill high-skilled operational posts.

In the next section we examine some of the manufacturing that is now being done in NC and discuss the implications of this 're-industrialization' for employment patterns. Although firms are now adding workers, the type of workers that they are seeking need skill levels that are qualitatively different from those hitherto possessed by traditional manufacturing workers. This has often resulted in a skill shortage/skill mismatch, the essence of which is discussed in detail as it pertains to NC.

Manufacturing rebirth

Some of the reasons for the recent growth of manufacturing in NC, and more generally other southern states, are similar to those that explained the take-off of southern industrialization in the 1960s onwards (Wright, 1986) – low wages, low taxes, low rates of unionization and a strong pro-business culture. Regional development initiatives and state and local incentive packages were instrumental in bringing businesses to the region during that period. North Carolina, like most states, has incentive packages designed to attract firms that will generate employment growth, either directly via hiring or indirectly through the use of required subcontractors. For example, the One North Carolina Fund is designed to 'help recruit and expand quality jobs in high value-added, knowledge driven industries' as well as those deemed vital to a healthy economy (see www.thrivenc.com. incentives/financial/discretionary-programs/one-north-carolina-fund). A similar state incentive fund is the Job Development Investment Grant (JDIG), which provides sustained annual grants to new and expanding businesses. Seen as engines of economic growth and ancillary job creation, such incentives are designed to exploit the comparative advantage of the region vis-à-vis higher-cost rust belt states.

North Carolina continues to attract new firms on the basis of low labour costs, good transportation resources and proximity to the eastern seaboard and southern sunbelt market. The following examples are an indication of the recent job growth in the state.

In October 2013, Owens Corning, a market-leading innovator in glass fibre technology and glass fibre materials announced that it will build a new advanced manufacturing operation in Gaston County, creating 110 new jobs. A Job Development Investment Grant from the state was provided based on proof of job creation and other performance indicators. In the same month, Trelleborg Coated Systems US, Inc., a leading manufacturer of engineered polymer solutions, announced the addition of 76 new jobs and a $10.6 million investment at its plant in Rutherfordton over the next four years. A grant of $440,000 from the state's One North Carolina Fund, again contingent upon proof of job creation, made the project possible. A figure of $300,000 from that same fund was made to Greenheck Fan Corporation in October for the establishment of a new manufacturing operation adjacent to an existing one in King's Mountain. The expectation is that 102 jobs will be created over three years. In each case the state argues that the grants are designed to attract business projects that will stimulate economic activity and create new jobs in the state. In turn, the companies state their satisfaction with the business climate, workforce training facilities and geographic proximity to key eastern US markets.

But existing companies that produce household goods (commercial appliances and consumer household products), HVAC equipment, and precision instruments and equipment have also flourished. Recent growth in segments of the textile and apparel industry, particularly yarn and fibre mills, and furniture makers is somewhat remarkable since these were sectors that lost most jobs during the 1990s. In other words, growth is not just consigned to 'new' manufacturing industries; more traditional sectors have also been able to capitalize upon rising production costs in Asian countries as well as domestic production innovations.

Textiles and apparel

Unifi Inc. was traditionally a high-volume supplier of commodity polyester and nylon filament yarns to the European and US textile sector. However, rising production costs forced it to outsource much of its production to Asia and Latin America by the late 1990s. During this period the firm restructured and downsized and in 2011 its Yadkinville plant was reorganized and a flexible modular manufacturing system was introduced. The plant now focuses upon specialist products such as its 'Repreve' brand which uses recycled goods, gaining economies of scale operations for recycled products. By reinventing itself as a technologically advanced textile manufacturer, it focuses upon proprietary and high-value-added niche products, thus avoiding cost-based competition with lower-cost overseas manufacturers. To realize this, Unifi put its workers through an extensive company re-training programme since the new manufacturing system was structured around interdependent team systems with multi-skilled operators. Wages and benefits are now higher than in the past and greater than prevailing local labour market averages.

Not too far from Unifi's facility is another specialty textile manufacturer, Burlington Technologies, Inc. Following receipt of a contract with the US military, in 2012 the company upgraded its production capability with a focus upon niche products. It has added 110 jobs and is spending $725,000 on capital investments during a three-year transition period. It has received $120,000 from the One North Carolina fund to help upgrade the skill sets of its existing workers and provide training for the new hires. In this instance a combination of productivity-enhancing capital investments has been combined with new job additions.

Finally, a Canadian company that had closed two of its NC plants in 2007 and eliminated 520 jobs in Mt Airy, NC, confirmed that it would be expanding its Mocksville yarn-spinning production capacity for new and existing products and creating at least 501 jobs in the state (Craver, 2013). Montreal-based Gildan Activewear Inc. is a leading supplier of quality branded basic apparel such as T-shirts, socks and underwear. Unlike the aforementioned textile companies which now pursue niche differentiation strategies, Gildan's expansion is a result of capacity constraints that have caused lost sales in commodity products such as socks and underwear. It has also expressed growing concern with lost production in overseas factories due to unreliable power sources and quality control problems. By locating more production in Davie County, NC, it hopes to leverage its vertical manufacturing to support a series of product growth areas, changing from a strategy of 'filling retail capacity to building additional capacity' (op cit: 10). It anticipates capitalizing upon lower cotton prices, improved capacity utilization and reconfigured distribution channels to drive business growth. The company has been awarded \$3.49 million over 12 years from JDIG as well as \$1 million from the Golden Leaf Foundation (a fund set up in 1999 to receive half of the funds set aside for North Carolina following the master settlement agreement with cigarette manufacturers) to buy machinery and equipment. For the workers, average wages of \$32,270 will be above those of the county average of \$28,028.

Furniture

NC, particularly the area in the central part of the state, has traditionally been a centre for furniture manufacturing for over a century. Abundant hard wood supplies from nearby forests, good railway and transportation infrastructure, and a plentiful supply of cheap labour contributed to the growth of the industry (Cater, 2005). The city of High Point (the site of an annual furniture trade fair that attracts buyers from around the world and earned the city the nickname 'furniture capital of the world') provided a focal point for the industry in the state and wider region. Employment during the 1980s averaged 84,578 jobs but, since then, a dramatic surge in imports (mainly from China) has resulted in factory consolidation and closings and continued reductions in the workforce (Nwagbara *et al.*, 2002; see also Figure 5.3). Between 2005 and 2010 25,746 jobs were lost and the industry now has about a half of the workers than at its height several decades ago.

In the last few years, however, rising wages rates in China and the lack of concomitant productivity increases amongst their workers, plus lower domestic energy costs in the USA, combined with new production systems by local US manufacturers to revitalize the industry. As with textiles, many existing firms have experimented in recent years with new manufacturing systems such as lean production that are designed to reduce production time, speed product throughput, minimize wastage and improve productivity (Russell, 2006). Similarly, they have focused upon niche areas that provide higher-value-added products and sought to leverage downstream links in the value chain, particularly in the areas of design and customized products of which their local retail market knowledge is crucial (www.soc.duke.edu/nc_globaleconomy). Capital investments mean that the industry now uses more technology; companies are more flexible, willing to customize and have smaller production runs; and workers are multi-skilled and

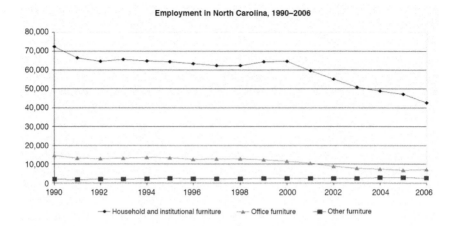

Figure 5.3 Furniture industry employment trends

Source: Employment Security Commission of North Carolina, 2007

able to understand complex technical systems. Many of the new jobs require critical thinking rather than mere knowledge of how to run a conveyor belt.

In 1995 Bruce Cochrane sold his family's 91-year-old furniture company, Cochrane Furniture, based in Lincolnton, NC because it was no longer competitive with imported products. After subsequently working in China as a consultant to the furniture industry, helping clients locate plants there, in 2011 he bought back the same plant and in December opened Lincolnton Furniture Company, manufacturing kitchen and bedroom furniture. By the following year he employed 130 workers. He had increasingly found that the rising cost of labour and raw materials in China was stymying many of the suppliers there, resulting in factories being forced to close and deliveries cancelled. With product demand in China increasing, many firms are facing a dilemma of how and where to meet that demand. As one commentator noted, in his analysis of a growing Chinese middle class who are now eager consumers of furniture,

> In some sense, the growth and demand in China is creating the need to build a new plant. Then, when I create a new plant, do I build that plant in China – or do I repurpose the plant in China and put another one in the U.S.? So it's really, in essence, the growth of China is allowing some of the plants to come back to the U.S.
>
> (www.npr.org/2011/10/27/141661114/back-from-china-furniture-maker-returns-to-n-c)

In nearby Hickory, 101-year-old Hickory Chair Company just added 86 new workers following three years of increased sales. President Jay Reardon attributes much of the success to a better utilization of employees as well as scrupulous attention to savings in areas such as energy usage (*Charlotte Observer*, January 4th, 2013). By soliciting input from workers, the company has improved work-flow, adjusted patterns to reduce wastage and concentrated on building quality into every facet of production ('building it right the first time'). He argues that well-motivated employees are his biggest asset and that his use of worker/super-visor teams to gather and implement new ideas has generated enthusiasm and commitment amongst his workers. Keeping labour costs (approximately $14–$18 an hour depending on tasks) to 18–22 per cent of total product costs and focusing upon customized products and high-end furniture means that he doesn't have to compete with low-cost, mass-produced manufacturers from Asia.

These examples are indicative of a trend where companies are finding it possible to relocate or expand production within the state, providing they can identify specific product ranges that are commensurate with the extra design, quality and customized demands that high-value-added products entail. Those that do this have embraced an operational model in which flexibility, efficiency and customization permit small, high-quality production runs. In turn this has been predicated upon investments in both high-tech machinery and, crucially, worker training.

According to many employers, however, it is the latter that has recently emerged as problematic. Like several of its manufacturing counterparts, the furni-ture industry faces a skill shortage dilemma. Its existing workers are older (28 per cent are 55 and older, compared to 19 per cent ten years ago), with many facing retirement. Newer workers are no longer being attracted to the industry, having seen successive generations of relatives being laid off, and the ones that are frequently lack the appropriate new skill sets. This is particularly problematic for the most skilled categories such as upholsters and sewers, who are in short supply. Furthermore many of the newer jobs in this more specialized industry require more cognitive advanced skill sets than in the past. As one furniture industry executive noted, 'the field needs techno-savvy young people who enjoy working with their hands. We use tools that are pretty space age. It is not just working in a dark, dusty shop' (quoted in *Charlotte Observer*, January 4th, 2013). Running complicated machines and being able to adapt to a flexible workplace require skill sets that are different to those associated with traditional, almost Fordist assembly lines.

New jobs, new workers, new skills?

The above examples suggest that the new manufacturing sector comprises firms that are leaner, frequently niche producers, who have focused upon value-added products to remain cost and quality competitive. In this technologically trans-formed workplace, worker skills have been upgraded, but this has come at a time when younger workers often lack the requisite skills or a desire to enter a sector that is mistakenly characterized by dirty, physical labour. With up to 40 per cent of Americans still possessing a high school diploma or less, the challenge remains how to instill a skill acquisition culture to meet future employer demands in a post-Fordist workplace. The alternative is for employers to increasingly assume the task of providing extensive worker training – something which in the past they have been reluctant to do (Green, 2013).

In Charlotte, Siemens faced skill shortages problems as it embarked upon a hiring programme. As part of a $135 million expansion at its manufacturing facility, it planned to add 800 skilled jobs (mostly machinists, welders and mechanical assemblers) that would require workers to possess rudimentary engineering capabilities and be able to interpret sophisticated blueprints. Despite numerous efforts, the company has thus far been able to hire only 100 new workers and has suspended its additional hiring. Instead, it has partnered with Central Piedmont Community College to help train workers, brought managers over from Germany to create training programmes similar to German apprenticeships and even offered to put new workers on the payroll whilst they are undergoing training. To ensure that it has the requisite skilled workers for its manufacturing expansion plans (tied to US demand for its light rail systems) in the future, the company is currently spending $165,000 an apprentice for a three-year mechatronics training programme (www.newsobserver.com/2013/10/11).

It appears that some other Charlotte-based companies have taken note. Carolina CAT, a heavy equipment manufacturer, recently adopted a similar technician-training programme whereby it sponsors students for 12 months at the local Community College followed by eight weeks of on-the-job training (Furhman, 2012). Under this system, all the workers who complete the training are guaranteed jobs with the company.

Such training programmes have not been commonplace in the past, given the neo-liberal market employment model that is pervasive in the USA, in which workers are assumed to build their own human capital and are more mobile than their European counterparts (Crouch *et al.*, 1999). Companies have often been reluctant to provide too much training since it equips a worker with valuable skill sets that could be used to acquire alternative employment. The flexible labour markets and worker mobility that are highlights of the US employment system thus run counter to the long term 'skill-building' training programmes that are now seen as crucial for a revitalized and high-tech manufacturing sector. The absence of a broader commitment to skill upgrading means that firms such as those listed above might continue to be the exception rather than the norm (see Rubery and Grimshaw, 2003).

As more firms lament skill shortages, attention has shifted to ways of ameliorating the problem and a possible rethinking of company policies to training. In a 2012 report on the state of the NC workforce in which a broad set of firms from different sectors were surveyed, it was noted that changes in the state's economy and primary business sectors have resulted in a demand for different skill sets than those possessed by many current workers (North Carolina Association of Workforce Development Boards, 2012). Almost 40 per cent of the survey's respondents were from manufacturers (the biggest group, followed by health and community services at 12.3 per cent), who indicated 'general maintenance' as their primary skill shortage but also that 'soft' skills (interpersonal/communication) and 'knowledge based technical proficiency' skills were frequently lacking in job applicants. One is always somewhat wary when firms list soft skills, as this can sometimes be managerial code for not having the 'right attitudes and values, such as conscientiousness and reliability' (Green, 2013: 17). Frequently this follows a lament about the education system or even family values that are perceived as providing inadequate socialization for the world of work.

But it might also be a reflection of forced changes in a decades-long commitment to neo- Taylorist and Fordist work practices in which semi-skilled workers performed routinized tasks under enhanced supervision. Numerous studies have pointed to de-skilling processes within firms following an earlier introduction of new technology because it was seen as a way of asserting greater workplace control (Taplin, 1992; Vallas, 2012). The new workplace, however, is one where broader competencies are required and workers are expected to have initiative and the technical background to engage in basic problem solving. The old system was predicated upon mass production and economies of scale in which firms primarily competed on the basis of cost. Now, as the above cases indicate, to be successful requires niche market strategies in which quality and sophistication are built into a higher-value-added finished product. This lessens the impact of higher unit labour costs, especially when combined with higher rates of worker productivity. But this is a dramatic departure from old practices, and not surprisingly labour force cultural norms will take time to adapt, since skill is socially constructed (Green, 2013).

Aside from general discussions about skill formation in a society, in which education is seen as playing an important role in general learning, firms are now beginning to talk more seriously about formal and informal training for skills specific to their particular work requirements. Employers are saying that there is a need for standardized work-readiness skills training and certification (North Carolina Association of Workforce Development Boards, 2012) and, even though many now conduct in-house training, they increasingly look to local Community Colleges as resources and even partners.

It is difficult to precisely enumerate the amount of money that specific firms spend on skill formation, but since most now talk enthusiastically about vocational training and have restructured around a technologically transformed workplace, anecdotally at least it appears to have grown. The same can be said about wages, where one is forced to rely upon aggregate measures. For example, 2012 mean hourly wages of production workers in Charlotte (where one-quarter of the state's manufacturing employment is located) were $16.01, which was higher than many other southern cities but lower than the national average of $16.45. Individual firm claims about average payment levels for new skilled workers vary considerably, from $14.13 an hour for sewing machine operators and $18.09 for upholsterers in a furniture plant, to $22 an hour in an advanced textile plant.

Placing a precise dollar value on skill is fraught with difficulty, reflecting variables from marginal costs to product demand in a firm at any one time. Competitive market rates for labour are generally assumed to be the best barometer for determining wages, but once one adds training programmes (in-house or paid for at a Community College), the calculations are more uncertain. There are uncertainties that surround the acquisition of skill, both from the individual who might make the investment based upon perceptions of what the future labour market looks like, to employers who are never certain that those selected for training will manage the transition to high-skilled work. There remains concern amongst some managers about the potential loss of skilled workers following training as they seek higher-paying jobs elsewhere. This can be an incentive to pay higher hourly wages, provide a workplace environment that sustains worker commitment (high-involvement working practices) and autonomy, and even focus more on training for job-specific skills that are less transferable. In tight labour markets, skilled workers might perceive that there is greater security in skilled work and thus become more compliant following training. If, however, employer concerns about turnover persist, then a more pessimistic attitude regarding training could potentially re-emerge and the opportunity to fundamentally redesign the workplace will be lost.

Conclusion

Forecasting the demise of US manufacturing has occurred before, following massive import penetration associated with Japan's rapid growth in the 1970s, only for the naysayers to be proved wrong when manufacturers re-engineered production, and inefficient firms were left to close (Sirkin *et al.*, 2011). The recent recession that followed a decade of liberalized trade and increased outsourcing – the ubiquitous globalization process – led to a resurgence of such sentiments. Yet again, segments of the manufacturing sector have begun to thrive. Capitalizing upon higher costs associated with overseas production, many firms have restructured to take advantage of comparatively low domestic costs and high-value-added niche markets. Southern states in the USA continue to be attractive for such endeavours because of their combination of production efficiency, geographic location and a business-friendly institutional environment. Just recently, Chinese and Indian textile firms announced their intention of setting up production in South and North Carolina to save money, as wages, energy and other costs rise in their respective home countries (McWhirter and McMahon, 2013).

What makes this recent renaissance notable, however, is the type of jobs that are being created. The old assembly line work in mass production manufacturing that employed semi-skilled workers is being replaced by skilled workers with more sophisticated training and educational background. In one respect this harkens back to the craft workers of old (nineteenth century) who were eventually replaced following standardization of production and whose skills proved increasingly irrelevant (and too costly). But this new labour force is smaller, and it remains questionable whether their wage and benefits packages will be commensurate with the new level of sophistication that they bring to the job. Technological innovations typically lead to employment growth. The growth of manufacturing in areas such as North Carolina reflects a continued interest by manufacturers in sourcing jobs where there are still significant wage advantages. It is this combination of demand for skilled work and its location in relatively low-cost areas that makes the discussion so intriguing.

A new, more subtle form of outsourcing is also occurring. Facing a skill shortage for these new jobs, many firms rely upon local Community Colleges to restructure their curriculum in order to provide education and training that is in line with the new work requirements. In some cases, firms provide some funding, but in many it is expected that the state picks up the tab. Eager to secure employment possibilities, most states are enthusiastic to adopt the new programmes; but, as was noted, predicting jobs of the future (and their requisite skill sets) is always fraught with uncertainty.

Continued high unemployment rates in NC are associated with the lack of jobs for workers with minimal skills. The recent growth of health care and a bio-tech sector in some regions is hailed as a partial solution to manufacturing decline, but the skills associated with such jobs are cognitively different to those of displaced, often older manufacturing workers. Retraining programmes continue to be funded by the state but not every laid-off worker has the capacity for high-skilled work. Jobs such as janitors, security personnel and restaurant workers are increasing, but these are frequently minimum-wage jobs, often part-time and with few if any benefits. Again, the skill mismatch debate resurfaces.

The resurgence of some manufacturing belies those who talk about a jobless recovery (Rattner, 2014). But it is now a more specialist sector, with qualitatively different sorts of jobs to that of the past. Whether this trend will continue and, if so, if it will eventually generate income gains for workers, remains an interesting question. What has occurred is notable in that it has reshaped the way we think about manufacturing work in the twenty-first century and whether the current form is post-Fordist or something completely different. It has also forced some to at least consider whether liberal market employment policies will continue to be viable in a high-skilled sector where education and training are crucial components of business success.

References

Al-Mahmoud, S.Z., Passariello, C. and Rana, P. 2013. 'The global garment trail: From Bangladesh to a mall near you', *Wall Street Journal*, May 4–5: A1,11.

Bluestone, B. and Harrison, B. 1982. *The Deindustrialization of America*. Basic Books: New York.

Cater, J.J. 2005. 'The rise of the furniture manufacturing industry in Western North Carolina and Virginia,' *Management Decision*, Vol. 43, No. 6: 906–924.

Cohen, S.S. and Zysman, J. 1987. *Manufacturing Matters*. Basic Books: New York.

Craver, R. 2013. 'Gildan to expand in Davie, Rowan', *Winston-Salem Journal*, September 10: A1.

Crouch, C., Finegold, D. and Sako, M. 1999. *Are Skills the Answer?* Oxford University Press: Oxford.

Davidson, A. 2013. 'Clotheslined', *New York Times Magazine*, May 19: 16, 18.

Dicken, P. 2007. *Global Shift* (2nd edition). Guilford Press: New York.

Employment Security Commission of North Carolina 2007. *Employment and Wage Data by Industry*. Raleigh, NC.

Green, F. 2013. *Skills and Skilled Work*. Oxford University Press: Oxford.

Furhmans, V. 2012. 'Germany's new export: Jobs training?', *Wall Street Journal*, June 14.

Hagerty, J.R. 2013, 'Whirlpool jobs return to US', *Wall Street Journal*, December 20: B4.

Hagerty, J. R. and Linebaugh, K. 2012. 'In US, a cheaper labor pool', *Wall Street Journal*, January 6.

Lowrey, A. 2013. 'Our economic pickle', *New York Times*, January 13: 5.

McWhirter, C. and McMahon, D. 2013. 'Spotted again in America: Textile jobs', *Wall Street Journal*, December 21.

North Carolina Association of Workforce Development Boards, 2012. *Closing the Gap. 2012 skills survey of North Carolina Employers*. Raleigh, NC.

North Carolina Commission on Workforce Development, 2007. *State of the North Carolina Workforce: An assessment of the state's labor force demand and supply, 2007–2017*. Raleigh, NC.

North Carolina Commission on Workforce Development, 2011. *State of the North Carolina Workforce: Preparing North Carolina's workforce and businesses for the global economy*. Raleigh, NC, June.

Nwagbara, U., Buehlmann, U. and Schuler, A. 2002. *The Impact of Globalization on North Carolina's Furniture Industries*. NC Department of Commerce, Raleigh, NC, December.

Porter, M. and Rivkin, J.W. 2012. 'The looming challenge to US competitiveness', *Harvard Business Review*, March: 54–62.

Rattner, S. 2014. 'The myth of industrial rebound', *New York Times*, January 26: 1.

Rubery, J. and Grimshaw, D. 2003. *The Organization of Employment*. Palgrave: Basingstoke.

Russell, T. 2006. 'Manufacturers getl,' *Furniture Today*, August 7: 47.

Sirkin, H.L., Zinser, M. and Hohner, D. 2011. *Made in America, Again. Why manufacturing will return to the US*. Boston Consulting Group: Boston, August.

Taplin, Ian M. 1992. 'Rising from the ashes: The deskilling debate and tobacco manufacturing', *Social Science Journal*, Vol. 29, No. 1: 87–106.

Taplin, Ian M. 1994. 'Recent manufacturing changes in the US apparel industry: The case of North Carolina', in Bonacich, E., Cheng, L., Chinchilla, N., Hamilton, N. and Ong, P. *Global Production*. Temple University Press: Philadelphia, pp. 328–344.

Taplin, Ian M. 1995. 'Flexible production, rigid jobs: Lessons from the clothing industry', *Work and Occupations*, Vol. 22, No. 4: 412–438.

Taplin, Ian M. 2006. 'Strategic change and organizational restructuring: How managers negotiate change initiatives', *Journal of International Management*, 12: 284–301.

US Bureau of Economic Analysis 2007. NC Employment, Department of Commerce: Washington, DC.

US Bureau of Economic Analysis 2011. Department of Commerce, Washington, DC.

US Bureau of Labor Statistics 2013. NC Employment, Washington, DC.

Vallas, S. 2012. *Work*. Polity: Cambridge.

Waters, R. 2011. 'Job-devouring technology confronts US workers', *Financial Times*, December 15.

Wright, G. 1986. *Old South, New South*. Basic Books: New York.

6 Institutional change, industrial logics, and internationalization

Growth of the auto and IT sectors in India

Anthony P. D'Costa

1 Introduction

Much has been written on India's recent high economic growth rates and its increasing industrial visibility in the world economy. The Indian information technology (IT) industry has become the poster child of the "flat world," exporting nearly 70 percent of its output, while the recent launching of the Tata Nano, the world's cheapest car, and the purchase of Jaguar Land Rover of the UK by Tata, seem to have finally signaled India's manufacturing as well as services prowess. These developments are a significant departure from the pre-1980s' India. Not only did India miss the manufacturing bus compared to many East Asian economies and now China, it was not known either for its technological strengths or major foreign acquisitions. Naturally, the question arises as to how this turnaround came about, given India's "soft state"—the presumption being that, in most Asian manufacturing hubs, state orchestration of industrial upgrading has been important (Amsden 1989; Evans 1995). Furthermore, unlike many East and Southeast Asian countries hit by the 1997 financial crisis, and the world economy as a whole with the 2008 global financial crisis, India seems to have been unscathed by both.

The ability to shield the economy from global financial shocks had previously been due to conservative monetary and financial polices, including with respect to managing current account deficits and the capital account, and regulating foreign short-term portfolio inflows.[1] More importantly, limited exposure of its industrial sectors to the global market subjected the Indian economy less to the instability of the world economy. Limiting macroeconomic and economic exposure to the world economy has been both a conscious strategy as well as a result of the interactions of initial conditions, institutional change, and industry logics. Thus the colonial legacy, and subsequent state-led development that emphasized domestic production, had a bearing on industrial outcomes. Similarly, in the 1980s when gradual economic reforms were phased in, both the Indian auto and IT industries were poised to take advantage of the pre-existing state–business relationship and tertiary education infrastructure. The policy of calibrated internationalization, initially through joint ventures with multinationals at home and outsourcing partners abroad, contributed to India's global engagement. Lastly, specific industry logics such as market size and demand characteristics influenced the growth trajectories of these sectors, even in times when the rest of the world economy was reeling under the financial crisis.

A state-led approach established industrial capacity in critical sectors, but with high-cost, inefficient industries, making deregulation for competitiveness necessary. However, this market-driven explanation is only partly true, since only a handful of sectors such as auto, pharmaceuticals, telecom, and software exhibit dynamic global competitiveness. The vast majority of Indian manufacturing is comparatively behind when benchmarked against China or other Asian economies. The accompanying institutional change, whereby new rules governing the workings of the economy have been devised to unshackle businesses allegedly stymied by bureaucratic and rent-inducing regulations, have partly contributed to India's industrial competitiveness.[2] Much of the dynamism comes from the gradual deregulation that allowed capable industries established behind protection to adjust to a new competitive environment, including capturing new global economic opportunities. These high-growth sectors were also subject to substantial state intervention well before the major set of liberalizing reforms which came in 1991; and therefore these reforms, leading to reduced state intervention, cannot solely explain their transformation. There are other factors such as initial conditions, institutional change, and industry logics that have worked in favor of certain industries in India.

I argue that the turning point of industries and their subsequent internationalization is better understood if the interplay of initial conditions of these industries along with specific industrial logics that govern these sectors is considered. For example, by addressing the question "What were some of the characteristics of these industries and the policy context in which they were ensconced before they became globally visible?" we are likely to get a sense of the industry trajectory. This trajectory, when combined with industry logics that entail key techno-economic and policy-political parameters, including the size and type of market and industry structure for particular sectors, allows us to identify the key factors for change that are otherwise lost sight of in the shuffle of large shifts in institutional arrangements, particularly those that privilege the workings of the market.

The objective of this paper is to explain the rise of two visible sectors: the auto industry (mainly the passenger car segment) and the IT industry (mainly software services). These are very different sectors, with different histories and policy environments. One of them represents a quintessential consumer durable, rooted firmly in high-value manufacturing and the other a high-value tradable service, which is dependent on the supply of technical professionals. The issue that India is doing well in high-value industries despite high levels of poverty and abundant availability of unskilled workers is important but it is not the focus of this discussion. Rather, the purpose here is to show how these industries, characterized by specific initial conditions and industry logics, have evolved into dynamic sectors.

My method for tracking industrial change in India is twofold. The first is to demonstrate the importance of the analytical approach taken, which in this instance is an evolutionary method. While this is not a novel approach, the marrying of industry logics with initial conditions and institutional change can be seen as a contribution to how we assess industrial performance. It suggests that, given certain preconditions, policy shifts could catalyze further change. In other words, the contextual dimensions of an industrial trajectory are key to understanding change, including those that are specific to India. The second is to show the critical role of the state in not only shielding these industries from the world economy but also taking advantage of it by reconstituting itself to gradually work with the market. Businesses have responded to an increased economic space through particular forms of self-regulation at the industry level and competitive strategies such as flexible business models.

The paper is divided into six sections. In Section 2 I develop a simple framework to capture industrial dynamism leading to internationalization. Section 3 identifies the key initial conditions of the case industries that set the stage for their expansion when the policy environment underwent radical change. In Section 4 I analyze the industry logics, such as markets, industry structures, and dominant business models adopted by these sectors. Section 5 presents a brief discussion of the internationalization of these sectors in terms of foreign direct investment (FDI), outsourcing arrangements, and diaspora. The final section concludes.

2 A framework for understanding industrial change

Economic growth is a complex phenomenon and yet it is typically explained by macroeconomic analysis of investment, interest rates, and other fiscal and monetary variables. Such an approach does not look into what lies behind growth. After all, growth is a sum total of all economic activities (never mind the measurement question and which economic activity is included or not) and there are clearly real actors making decisions about where and how much to invest. A well-functioning market system is expected to coordinate economic decisions through the forces of supply and demand. However, macroeconomic analysis is neat and pertains to national growth rates. It does not deal with how specific industries contribute to that growth rate, assuming that there must be some relationship between a high economic growth rate and high industrial growth rates. Since industries are a set of building blocks of a modern economy, it is equally important to understand how and why they emerge in particular institutional contexts.

The common line of argument is that, if the previous institutional arrangement was hampering growth, while changes in institutions resulted in higher growth rates, then institutional change is the explanatory variable (à la North 2010). Since India had deregulated and liberalized, it suggests that market dynamics explain the rise of these industries, given that the previous nationalist model was not particularly effective. Attractive as this interpretation seems, institutions are messy affairs since, aside from the formal rules and regulations that govern industries, there are many informal institutions, which in a developing country context could turn out to be significant. Besides, markets work best when there are corresponding institutional props (Rueschemeyer 2009: 162), and specifying growth-enhancing institutions is not easy. One institution that has received much attention in industrial change is the state (Amsden 1989). However, this calls for state capacity and coherence (Kohli 2007), which in turn rests on embedded autonomy (Evans 1995). The Indian state is more fractious, with a divided bureaucracy and clienteilism. An alternative explanation for industrial change is politics, whereby collective action problems are solved by "consultation," "credible commitments," and "monitoring" (Doner 2009: 16). This is in effect a form of self-regulation of an industrial sector within the broader state–business institutional arrangement. The political position of the sector, the lobbies that represent it, and the sector's strategic importance to the state are also important for understanding varying performances of industries in the same economy.

All of these approaches are useful to understanding industrial change in developing countries. However, I take a different, multi-layered approach because the two industries under investigation are sufficiently different in origin, characteristics, and trajectory. Both have experienced the wider institutional change but are driven by different markets, namely, domestic (auto) and international (software). Rather than rely solely on policy regime change as the explanation of industry turning points, which is common to both scholarly analysis and popular reporting, I take an evolutionary view so that the interplay of the initial conditions, industry logics, and business responses can be analyzed in the context of changing policy regimes.

The basic argument is that in order to capture the turning point and thus internationalization of the two industries we need to identify some of the key initial conditions, such as the status of the industry before the turning point to identify the trajectory. Next the industry logics, which encompass both the policy-political and the techno-economic dimensions, must be linked to the evolution of the industry. Industry logics suggest that, once the industry is already on an expansionary path, it compels institutional change. As Evans (1995: 116) shows, the Department of Electronics in India adopted a promotional role when the industry was already on the threshold of expansion. To put it more strongly, purposive social and economic forces can reinforce the move upward of an already favorable trajectory of a sector, in a virtuous cycle. Thus, as firms in both the auto and IT sectors attained economies of scale, reflected in the industry structure, their market gains became more expansive. Figure 6.1 presents this evolutionary dynamic.

The initial conditions as reflected in the status of the two industries prior to the turning point include three dimensions: import substitution industrialization (ISI) regime, middle-class growth, and the exhaustion of ISI. These initial conditions were applicable for both the auto and IT sectors. India's post-independence period until the early 1980s was characterized by an ISI regime that promoted self-reliance through domestic manufacturing by the state or the private sector (D'Costa 1995). This was a nationalist strategy that aimed to shield the Indian economy from external shocks. Policy instruments included tariff protection, local content rules, industrial licensing and production capacity controls, subsidies, and entry barriers for multinationals. While many of these policies

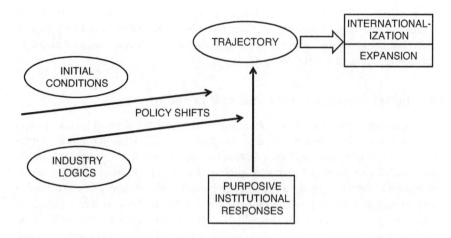

Figure 6.1 Evolutionary dynamics of industrial sectors

enabled a former colony to industrialize, there were several shortcomings. Since many policies once enacted had their own momentum, and since some resulted from political expediency rather than economic merit, Indian industries became high cost and fell behind the international technological frontier. Consequently, pressures for economic reforms, namely deregulation and liberalization, mounted feebly from the late 1970s and more strongly in the mid-1980s. However, it was only the 1991 reforms that radically broke with this nationalist policy framework.

It is evident that unfavorable industrial performance induced a feedback mechanism for reforms later. The reforms were introduced at a time when the Indian capitalist class had matured considerably, there was a sizeable middle class, and India had created a large tertiary educational infrastructure, especially for technical professionals. These developments impacted the development of the auto and IT industries. These are not causal factors (although consistent with endogenous growth theory), but preconditions, which placed the two sectors on a favorable trajectory. Once the contingency of expansionary polices was introduced in the context of a large, growing middle class, which included a large pool of technical professionals, the industry logics propelled the industries further and placed them on a dynamic path.

By relating these initial conditions to industry logics such as best industry practices and business models adopted under shifting policy regimes, we witness the foundations for internationalization and competitiveness. Naturally, there have been idiosyncratic factors such as ad hoc policies, sectoral autonomy, and historical accidents that also contributed favorably to the industrial change; but the element that propelled the sectors once they were on a trajectory was their ability to flexibly adjust to changing market conditions. Demand, as evidenced by high economic growth and a growing middle class, is certainly an important dimension of market dynamics. However, there were supply factors in the form of business models that also contributed to industry expansion. We turn to both supply and demand factors for the two sectors below in the context of initial conditions.

3   Initial conditions in the auto and IT industries

In this section I first present a brief evolutionary understanding of the two Indian sectors, auto and IT. I begin with the immediate post-independence period to indicate the regulatory environment of the auto industry and the consequent industry structure. I also carry out a similar exercise for the IT industry, with the caveat that the IT industry emerged much later globally. Nevertheless, the electronics and computer industry predating the dynamic software sector was also regulated with the intent to create a domestically oriented, self-reliant economy. One of the effects of a nationalist strategy that unwittingly aided India's industrial performance later, especially the IT industry, was the creation of a technical education infrastructure. Next, I analyze the significance of the Indian middle class as it relates to economic demand from a consumption point of view and the supply of technical professionals for the IT industry. Human resources are a critical input for global participation in terms of scale and skills, both of which were met through the relentless pursuit of professional education by the middle class.

3.1   The import substitution industrialization regime

In 1949 the government of India banned the import of completely built vehicles and, since 1953 under the aegis of the Tariff Commission, has refused permission to Indian manufacturers to assemble imported vehicles without increasing local content. This was labelled the "phased manufacturing program." With this measure the government reduced the number of assembly firms from 12 to five (Kathuria 1990: 2) and in the end was left with two main producers: Hindustan Motors of the Birla family and Premier Automobiles of the Walchand Group.

Like other industries, the auto industry, especially passenger cars, was highly regulated. First, cars were seen as a luxury product. They were taxed heavily and output was restricted through production licenses. Second, production was limited to a single model. Third, technology imports were restricted, in part to encourage domestic capabilities and in part to stem the outflows of scarce foreign exchange. In fact, the auto industry came under particular scrutiny because of India's persistent balance of payments problem. The regulatory environment at times became confusing (see Bhagwati 1993). On the one hand, the state imposed price controls (until 1975) to pre-empt monopoly pricing; and on the other, it restricted output. Production, as noted, was limited to a single model. The net result was a high-cost, technologically obsolete industry.

Auto component manufacturing was reserved for small-scale industry, but in the absence of economies of scale, the industry remained highly fragmented and technologically underdeveloped. After the energy crisis of the early 1970s, policy changed in favor of unlimited production of commercial vehicles and two-wheelers (Pinglé 1999: 99). Government priorities remained steady with the promotion of buses and trucks but also loosened for the passenger car segment to cater to the incipient class of consumers. Gradually, policies shifted in favor of liberalization. The restrictions on imports of capital equipment (technology) for replacement were relaxed with the condition that net foreign exchange outflow was zero. This had two effects, conserving foreign exchange and pushing firms to export. The bureaucratic process of permits for net imports was significantly simplified and automatic capacity expansion permitted for commercial vehicles and two-wheelers. Later, limited liberalization was also extended to the components industry. These somewhat piecemeal reforms had a cumulative effect of loosening the tight grip of the state on the industry, allowing scale economies on a limited basis and upgrading of technology in the non-car segment.

The initial conditions for the growth of the IT industry have been somewhat different. There was no software sector to speak of in the heyday of the ISI regime. What emerged, after the electronics revolution, was the computer or hardware sector in the 1960s. Multinationals from the USA and Western Europe dominated global markets through the leasing of hardware, serviced by home-grown technical professionals. However, due to defense industry needs, there was some strategic intent on the part of the Indian government to facilitate the development of the electronics/computer industry. For example, administratively, the Indian government established the Department of Electronics to foster an electronics industry (Evans 1995; Sridharan 1996; Rajaraman 2102). Related to defense, many public sector high-technology research organizations were set up; in addition to the Indian Air Force headquarters, the Indian Space Research organization and Hindustan Aeronautics Ltd, were created, which predate the growth of the IT industry (D'Costa 2009). Over time government-sponsored research in astrophysics, space, artificial intelligence, basic sciences, microwaves, power, biological sciences, and mathematical modeling and computer simulation was also established. More directly, engineering and science colleges such as the Indian Institute of Science, and IT training centers, and critical satellite infrastructure in centrally organized software parks, have served the industry well (Heitzman 2004: 222–229). While this policy created the basis for an electronics industry, it was inadequate to be internationally competitive with the likes of Taiwan, Singapore, or South Korea. At the same time, self-reliance remained the driving force behind the IT industry, which restricted the imports of technology (hardware in this case).

In the 1970s the state set up the National Informatics Center, the Computer Maintenance Corporation (CMC), the National Center for Software Development and Computing Technology, and regional computer centers. In 1977, Indira Gandhi's election defeat brought a coalition government to power. By this time, creeping reforms were in process. However, under the *Foreign Exchange Regulation Act* multinationals had to reduce their equity to minority shareholding. Both computer firms IBM and ICL (of the UK) declined and left the country (Encarnation 1989). For these firms, exiting was not a major loss since they were not producing anything special for the Indian market, which was small in any case. More importantly, globally the IT industry was on the threshold of a massive boom, with electronics manufacturing as well as markets gravitating toward East and Southeast Asia. However, the vacuum created by the departure of the hard-ware multinationals in India prompted the government to create the Computer Maintenance Corporation, a public sector firm designed to service the hardware sector in India (Grieco 1984). The company became an important training ground with tens of thousands of engineers on its payroll. With restrictions in technology imports, Indian firms were compelled to devote highly skilled professionals to hardware maintenance and rudimentary development. Inadvertently this created a large reservoir for potential mobilization of professionals in the IT industry once the need for a national hardware industry dissipated.

3.2   Middle-class formation and the exhaustion of ISI

The results of ISI as practiced in India are well known. Due to a plethora of regulations, often conflicting, businesses were inefficient, adopted non-Schumpeterian responses to profit opportunities, produced high-cost and poor quality products, and contributed to an economy of shortages. In the midst of these challenges India's urban consuming classes (namely the white collar profes-sionals, corporate employees, small business owners, organized labor, and public sector employees) experienced a rise in their expectations. The demonstration effect was strong because of the return of unskilled and skilled migrant workers and skilled professionals from the Middle East in the 1970s. In tandem, the economic rise of East and Southeast Asia illustrated the deep economic chasm in terms of consumption that existed between India and other parts of Asia, which erstwhile were poorer than India. Neither the quality of domestic products nor level of prices could satisfy the aspirational consumers. A long queue existed to obtain cars or scooters in India, two quintessentially middle-class products. Table 6.1 shows the rise of India's middle classes.[3]

There were two interrelated developments that led to a severe disjuncture, thus prompting economic reforms. The first was the rise of the middle (consuming) classes and the increasing access to tertiary education (a middle-class domain). The second was that these developments took place in an economic system which neither provided the goods corresponding to what was possible (the demonstra-tion effect) nor generated employment that matched the training received by many of India's youth (as reflected in emigration of professionals or disguised employment). Reforms were clearly in the offing to address both consumption and employment. Pertinent to the auto case, with pent-up middle-class demand, the lagging auto industry was given a boost with an unusual chain of events in the context of an emerging consuming class. In1982 the government of India created a public sector company, Maruti Udyog Limited (MUL), as a joint venture with Suzuki Motors Corporation (SMC) of Japan.[4] The government owned 80 percent of the equity. MUL emerged from a failed enterprise started by Sanjay Gandhi, the second son of prime minister Indira Gandhi. The high price and shortages of

Table 6.1 Three estimates of the size of the Indian middle class (based on household income)

NCAER (1989–90) Income category	NCAER (1989–90) Income group	% of population	ORG (1989–90) Income group	% of population	NCAER (1995–96) Income group*	Rural (m hh)	Urban (m hh)	% of hh
Low middle	Rs.12–25,000 ($721–1,502)	27.0	–		Rs.16,001–22,000 ($478–658)	7.1	36.9	26.7
Middle income	Rs. 25–40,000 ($1,502–2,403)	10.0	Rs.18,012–30,000 $(1,082–1,802)	13.4	Rs.22,001–45,000 ($478–1,345)	16.8	37.3	32.8
Upper middle	Rs. 40–56,000 ($2,403–3,364)	2.6	30,012–48,000 ($1,803–2,883)	4.7	Rs.45,001–215,000 ($1,345–6,428)	16.6	15.9	19.7
High income	Rs. 56,000+ ($3,364+)	1.4	48,000+ ($2,883)	2.0	Rs.215,000 + ($6,428+)	0.8	0.4	0.7

Notes: NCAER = National Council of Applied Economic Research, New Delhi; ORG = Operations Research Group, New Delhi.
 * 1994–95 prices
 The exchange rates used to convert rupee incomes in 1989–90 and 1995–96 were Rs.16.65 and Rs.33.45 per US dollar respectively.
 m = millions; hh = households
 India's per capita income (pci) in 2000 was $460; the purchasing power parity (ppp) pci was $2,390
 Based on ppp, Indiastop.com estimates were as follows: upper-middle-class, mid-middle-class, and lower-middle-class households at 40 million (with $600,000 a year), 15 million (with $20,000 a year), and 110 million (with unknown income, especially for the affluent rural bourgeoisie), respectively.

Sources: Based on sources reported in D'Costa (2005: 64)

automobiles in the 1970s inspired Sanjay Gandhi to set up a "people's car" project. However, not a single vehicle was manufactured and, with the re-election of Indira Gandhi in 1980, the enterprise was nationalized in 1981 (Pinglé 1999:107), and for the first time the state became an investor in a car project.[5] Suzuki was chosen mainly for its specialization in fuel-efficient small cars.

After the formation of MUL, the government in the mid-1980s relaxed some of the regulations and allowed other Japanese firms to form joint ventures in the commercial vehicle segment (see D'Costa 2005). However, no new passenger car manufacturer was permitted until the early 1990s. This was largely to protect MUL and arguably save foreign exchange by restricting additional vehicle producers. The commercial vehicle joint ventures were set up mainly to enter the Indian market even though the market for the smaller Japanese-designed trucks and buses was not appropriate for the Indian infrastructure conditions at that time. In fact, all of the Japanese joint ventures in the commercial vehicle segment largely failed due to lack of economies of scale and a formidable competitor, Tata Motors (D'Costa 2005). Tata had solid in-house engineering capabilities, allowing it to produce advanced machine tools at 25 percent of Japanese costs (Kathuria 1996: 215), abetted by its long familiarity with the Indian business environment. The low volumes of Japanese joint ventures could not keep up with Tata's much larger output of commercial vehicles, capturing over 50 percent of the market in which the Japanese were operating within a few years and more than 100 percent of the same market segment in 1997.

The growth in the middle class was also accompanied by increasing enrollments in an expanding tertiary education system (Table 6.2). Aside from the fact that India's education policy was highly skewed in favor of tertiary education, mainly to act as a source for trained workers for ISI, the limited places relative to demand generated intense competition amongst India's middle-class students. Since education was an important means for social mobility, it was highly sought after, even as the ISI regime was unable to generate either the volume or the type of jobs that the education system was generating. India's middle classes ensured that access to merit-based tertiary education remained their domain. Generally those who were educated in "missionary"/"convent" English medium schools and had a bachelor's degree from one of the better colleges were hired by a handful of multinationals, big business, and the public sector. With underemployment of the highly skilled professionals, emigration of engineers and doctors to the West became an important response.

India's IT visibility was due to middle-class dominance of technical education. This is especially evident in southern India. Bangalore, in the southern state of Karnataka, is home to several prominent public sector research organizations. The spread of tertiary education generally and in the greater Bangalore area contributed to the development of the IT industry. Though the state of Karnataka has only 5 percent of India's population, it has nearly 15 percent of its higher education enrollments.[6] It has 12 percent of the country's degree colleges under universities granting technical degrees and 15 percent of the country's diploma-granting polytechnics (Okada 2004: 298). In 2000–01 Karnataka had over 100

Table 6.2 Number of recognized educational institutions in India

Years	Primary	Upper primary	High school*	Colleges for general education	Colleges for professional education	Universities**
1950–51	209,671	13,596	7,416	370	208	27
1955–56	278,135	21,730	10,838	466	218	31
1960–61	330,399	49,663	17,329	967	852	45
1965–66	391,064	75,798	27,614	1,536	770	64
1970–71	408,378	90,621	37,051	2,285	992	82
1975–76	454,270	106,571	43,054	3,667	3,276	101
1980–81	494,503	118,555	51,573	3,421	3,542	110
1985–86	528,872	134,846	65,837	4,067	1,533	126
1990–91	560,935	151,456	79,796	4,862	886	184
1991–92	566,744	155,926	82,576	5,058	950	196
1992–93	571,248	158,498	84,608	5,334	989	207
1993–94	570,455	162,804	89,226	5,639	1,125	213
1994–95	586,810	168,772	94,946	6,089	1,230	219
1995–96	593,410	174,145	99,274	6,569	1,354	226
1996–97	603,646	180,293	103,241	6,759	1,770	228
1997–98	619,222	185,961	107,140	7,199	2,075	229
1998–99	628,994	193,093	112,050	7,494	2,113	237
1999–00	641,695	198,004	116,820	7,782	2,124	244
2000–01	638,738	206,269	126,047	7,929	2,223	254
2001–02	664,041	219,626	133,492	8,737	2,409	272
2002–03	651,382	245,274	137,207	9,166	2,610	304
2003–04	712,239	262,286	145,962	9,427	2,751	304
2004–05	767,520	274,731	152,049	10,377	3,201	407
2005–06	771,082	288,199	154,032	11,549	4,991	350
2006–07	756,950	300,008	165,087	11,458	7,024	371
2007–08	785,950	320,354	171,862	11,458	7,024	677***

Notes: * Includes higher secondary and intermediate and pre-junior colleges.
  ** Also includes deemed universities and institutions of national importance.
  *** The number is for 2014 and includes central, state, private, and deemed universities and institutions of national importance.
  1. Professional education includes engineering, technology, and architecture, medical (allopathy/ayurvedic/homopathy/unani/nursing/pharmacy, etc.), and teacher training colleges.
  2. Data on colleges for professional education for the years 1975–76, 1980–81 and 1985–86 includes institutions of post-matric courses.

Source: Reserve Bank of India; Ministry of Human Resource Development, Govt. of India (ON117)

engineering colleges, and nearly 12,000 students took IT-related courses (D'Costa 2009: 85). Visvesvaraya Technology University alone has 208 affiliated colleges, offering both Bachelor and Master of Engineering degrees with nearly 70,000 students.[7] Most students graduating from these colleges are proficient in English, basic engineering skills, mathematics, and programming. Similarly, Bangalore University has over 50 affiliated colleges located within Bangalore.[8] Although not a source of engineers, these colleges contribute to English-speaking science- and IT-proficient graduates. All in all, India's rise in the IT industry has to do with the supply of technical professionals, which the middle class dominates (D'Costa 2011a). The state of Karnataka and its capital city Bangalore have played no small part in the evolution of the IT industry.

In the mid-1980s, Rajiv Gandhi's government initiated significant economic reforms despite resistance from certain quarters of the bureaucracy, organized labor, and the private sector (see Kohli 1992: 322).[9] As the global IT industry began to take off, multinationals sought alternative production sites as a generalized strategy for lowering costs and tapping labor. East Asia emerged as a significant region for electronics manufacturing, with Taiwan, South Korea, Japan, and Singapore the leading producers. India was outside of this orbit. However, there was growing interest on the part of US multinationals to look into India as well for IT-related development work. The Indian government had already created space for the private sector after the departure of IBM (Evans 1995: 114). Rajiv Gandhi initiated deregulation for private business and liberalization of imports and foreign investments. In addition, the Indian diaspora was making its presence felt in Silicon Valley in the USA, and the interest of multinationals in India was once again ignited. In 1985, Texas Instruments from the USA came to India to explore the possibility of research and development. The successful completion of a small project led to more projects (see D'Costa 2000). Texas Instruments (India) from an initial foray into CAD tools moved into chip design in what has become its largest R&D center outside the USA. Its first major breakthrough came in 1998, 13 years after its inception in India, when it developed the "Ankur" digital signal-processing (DSP) chip. Other multinationals began to enter India in the 1980s, while several established Indian IT firms (TCS, Wipro, Patni, etc.), and new entrepreneurial firms such as Infosys and Mastek also emerged.

That the state played a role in creating industrial capacity is beyond doubt. What is not clear is why the trajectories of auto and IT took the form they did. The ISI strategy did not favor a competitive auto industry because of national priorities, heavy protection, bureaucratic management, and limited internal demand. What state intervention did do is alter the structure of the Indian economy, albeit gradually and contribute to a growing middle class. The ISI strategy required investments in tertiary technical education, which was the principal mode of social and economic mobility for the middle class. Thus, when industrial policies changed in the auto industry with a joint venture between the government of India and Suzuki Motors, the upward trajectory was inevitable because of preexisting latent demand (D'Costa 2005). With further reforms, Indian demand has been rising, which has attracted almost all multinational automotive firms.

The formation of joint ventures was a pragmatic strategy for multinationals to enter the Indian market as well as part of government policy designed to keep ownership partly in Indian hands. Some joint ventures are now 100 percent subsidiaries of multinationals such as GM's Halol facility in the state of Gujarat. Others such as Maruti-Suzuki and Toyota-Kirloskar are joint ventures with minority shareholding by Indian partners. Indian firms such as Tata have acquired internationally branded firms (Jaguar Land Rover), while Mahindra has set up joint ventures of its own in the USA and China to produce tractors. Similarly, the Indian IT industry began with multinational interest in sourcing software services. Over time the sector developed its own capabilities to cater to international markets. Not only are Indian firms exporting services on a large scale, they are also expanding abroad. In parallel fashion, multinationals are also investing in India to take advantage of the large pool of technical professionals. Thus, irrespective of the ownership status of these companies, they have collectively contributed to India's auto and IT industry internationalization.

4 Industry logics

Given the initial conditions of some deregulation, ad hoc protectionist state policies toward the auto joint-venture, and the concomitant growth of the Indian middle class and the availability of technical professionals, I demonstrate briefly how the two sectors' movement, already on an upward trajectory, gained momentum due to both purposive action (auto industry first and IT later) and by historical accidents (externally driven software services). The growth of the market and the industry structure is pertinent to this discussion since the monopoly position of MUL by all accounts should have led to clientelism. Instead MUL, by and large, can be credited with creating an entirely new industry in India with new products and, more importantly, with a new (Japanese-style) business model (see Bhargava 2010). This gave the firm and its suppliers the flexibility needed to compete in an expanding market even as the number of players increased substantially when the auto industry was completely liberalized in 1993.

The IT sector offers a different story, since the industry is largely about human capital. However, here too there was considerable input from the state, initially by creating a space for the private sector, incubating highly trained engineers in the public sector, and through the public educational and research infrastructure. Later the state contributed to IT infrastructure by creating a system of Software Technology Parks (established in 1991) and providing financial incentives (a ten-year tax holiday) to companies exporting all of their output. While the private sector became the lead player in the international outsourcing for the software services sector, the state as well as private engineering and technical colleges responded to the continuous global demand for such services. In other words, the markets, and thus institutions, adjusted to the trajectory of the sector and reinforced its expansionary path through its highly adaptive business models.

4.1 Markets and industry structure

As a first comer to the auto sector, meeting newly imposed fuel-efficiency standards, MUL secured a solid head start in a new line of business. It also had considerable help from the government. MUL's advantage in fuel efficiency over other Indian companies allowed it to import components at a reduced tariff— ranging from 40 percent in 1983–85 to 70 percent in 1991. Other companies allege that their import duty burden was 150 percent (D'Costa 2005). MUL also benefitted from imports of pre-cut steel, which was defined as a component with 40 percent duty, whereas steel sheets were interpreted as raw material with 150 percent duty. Pre-cut steel imports meant less waste, greater production efficiency, and higher profits for MUL. With effective elimination of a competitor, MUL's product mix, superior production technology, relatively large capacity, and government support in the initial years made MUL virtually unassailable. MUL captured 50 percent of the passenger car market by 1990 and has more or less retained that share today even as the total market has increased, by nearly fivefold since 2000 to 2.5 million units in 2010–11. Economies of scale allowed MUL to attain 88 percent local content for its principal vehicle within five years (Kathuria 1990: 33), and expanding output ensured its industry leadership. A consuming class accustomed to expensive and shoddy products found the Japanese vehicle aesthetically pleasing and economically attractive. MUL was the first to tap into the pent-up demand built up over the decades by a growing middle class.

Consequently, India's auto market expanded, as evident from output data, and simultaneously the industry structure was radically altered (Table 6.3). The incumbent firms were weakened but not completely eliminated. For example, HM (Hindustan Motors) in 2012 had only 0.16 percent of the market. However, Tata was a different story. It got into passenger cars and built up a market share of 11 percent and expanded abroad with branded products. The new fuel-efficiency norms were applicable only to *new* enterprises and *new* products. Established producers, such as HM and PAL, did not have to comply with fuel-efficiency standards for existing products, thereby averting plant shutdowns and labor strife in older firms. However, they were no match for the state-of-the-art production unit of MUL. As the industry became fully liberalized, more multinationals entered, with Hyundai of South Korea as the first 100 percent multinational subsidiary and others initially as joint ventures, such as General Motors with HM and Toyota with Kirloskar. MUL had already prepared the expansionary path with its own growth strategy but also built up a solid supplier industry, which also diffused to other areas of the country. Today MUL continues to control nearly 40 percent of the Indian car market. Chennai in the south with Hyundai (second largest producer in India, with 15 percent market share) and Ford is a highly competitive region with modern production facilities. Several Indian firms are now specializing in small cars for the world market, and several multinationals, especially from Japan, are using India as a platform for compact cars for third country markets.[10] The components sector has also become an exporter with nearly $10 billion in exports in 2013–14, albeit it remains a net importer of about $3 billion.[11]

The growth of the Indian information technology industry (which includes software and hardware) has been quite remarkable as well, exceeding 30 percent in most years. Even more remarkable has been the increase in India's software exports. Soon after the 1991 reforms the Indian software industry expanded rapidly, meeting the global demand for services with an abundant supply of high-skilled but low-cost professionals. In 1990, India's export of IT services was

Table 6.3 Changing market structure of car and utility vehicle production in India (1955–2001)

	HM		PAL		SMP		SAL		M&M		MUL		TELCO		Others##		Total
	# of units	Mkt. share	# of units	Mkt. share	# of units	Mkt. share	# of units	Mkt. share	# of units	Mkt. share	# of units	Mkt. share	# of units	Mkt. share	# of units	Mkt. share	
1955	4,874	37.9	3,581	27.8	1,526	12.0	–	–	2,864	22.3	–	–	–	–	–	–	12,865
1960	9,217	37.5	6,616	26.5	3,364	13.7	–	–	5,501	22.4	–	–	–	–	–	–	24,598
1970	22,703	51.0	12,054	27.0	448	1.0	–	–	9,334	21.0	–	–	–	–	–	–	44,539
1980	21,752	47.7	8,729	19.1	6	0.0	51	0.1	15,068	33.0	–	–	–	–	–	–	45,606
1990	26,204	12.0	42,737	19.5	–	–	924	0.4	32,706	15.0	116,194	53.1	*265	–	–	–	218,765
1997	24,059	5.0	14,169	2.9	–	–	–	–	69,277	14.3	349,780	72.0	6,302	1.3	22,545	4.6	486,132
2001	23,987	3.5	0	0	–	–	–	–	56,380	8.3	356,608	52.7	82,195	12.2	157,076	23.2	676,246

Notes: HM = Hindustan Motors, PAL = Premier Automobiles Ltd., SMP = Standard Motors Private Ltd., SAL = Sipani Automobiles Ltd., M&M = Mahindra and Mahindra, MUL = Maruti Udyog Ltd., TELCO = Tata Engineering and Locomotive Company

* for 1991

Others## include Daewoo, General Motors, PAL–Peugeot, Mercedes-Benz and for 1999 onward Hyundai and Fiat

By 2001, Daewoo, Fiat, PAL-Peugeot, and PAL had stopped operations

In 2001, Hyundai and Toyota had 57.3% and 18.1% of "Others" shares, respectively

Sources: Association of Indian Automobile Manufacturers (AIAM) and Automotive Components Manufacturers Association (ACMA) (various issues) in D'Costa (2005: 86)

$131.2 million, which by 2002 had reached $7.8 billion and now stands at over $50 billion. Software exports became an important foreign exchange earner for the country. Currently, such exports are nearly a quarter of all exports and the IT industry as a whole represents over 6 percent of gross domestic product (GDP) (D'Costa 2014: 171). In parallel fashion, most global information technology firms are present in India, undertaking a variety of development work for their in-house needs. However, today the largest IT firms operating out of India are Indian firms (Table 6.4).

The industry structure for the IT industry is different from the auto industry since entry barriers are low. Unlike the auto industry it is not capital intensive and relies on skilled professionals generally with technical backgrounds. As we have seen, tertiary enrollment has been growing in India as a response to increasing demand driven by exports. There are nearly 1,000 firms in the Indian IT sector, with many of them operating overseas. However, among these the top 20 firms control over 60 percent of the market (Rediff.com 2011). The top three firms are Indian, whereas 13 of the top 20 IT firms are multinationals and their subsidiaries. Although the top three firms have been always Indian, there has been an important shift in the industry in terms of ownership. In 1997–98, of the top 20 firms only five were foreign (D'Costa 2000: 156) and most recently in 2013–14, 19 of the top 20 were Indian firms (NASSCOM 2014). Thus, although the industry has many firms and most of them are Indian, multinationals have made big inroads

Table 6.4 The top eight Indian IT firms (rank)

1980–81	1985–86	1989–90	1994–95	1997–98	2007–08	2012–13
TCS	TCS	TCS	TCS	TCS	TCS	TCS
TUL**	TUL	TUL	TUL	Wipro	Infosys	Infosys
Computronics	PCS	COSL	Wipro	HCL Consulting	Wipro	Wipro
Shaw Wallace	Hinditron	Datamatics	Pentafour	Pentafour	Satyam	HCL Technology
Hinditron	Infosys	Texas Instruments*	Infosys	NIIT	HCL Technology	Tech Mahindra
Indicos Systems	Datamatics	Digital*	Silverline	Infosys	Tech Mahindra	iGate
ORG	DCM DP	PCS	Fujitsu*	Satyam	Patni	Mphasis
System	COSL*	Mahindra-BT**	Digital	Tata Infotech	Oracle Finance*	L&T Infotech

Notes: 1980–81 to 1994–95 rankings refer to top Indian IT exporters
    * foreign firms; ** TUL joint venture between Tata and Unisys, Mahindra-BT joint venture between Mahindra and British Telecom

Sources: Heeks (1996: 89) for data up to 1994–95, D'Costa (2000: 156) for 1997–98, and NASSCOM (2014) www.nasscom.in/industry-ranking for 2007–08 and 2012–13 (accessed 05/06/2014)

into India, just as Indian firms have opened development centers abroad. The reasons are straightforward: India is a growing market and India has the talent pool. The top firms also enjoy economies of scale as reflected in the number of employees. For example, the top firm, TCS, had 198,500 employees, followed by Infosys with nearly 140,000. These firms have been able to mobilize technical professionals for large-scale complex projects and gradually have taken over mission-critical services for foreign clients. This means that, even when overall demand goes down, firms will still need IT services from India to keep operating. The response of the economy to the growth of the IT industry has been to reinforce more growth with expansion in education (see Figure 6.2) and continue to reap the benefits of scale. Thus it is also not surprising that, despite the global financial crisis, the Indian software sector continues to expand.

4.2 Flexible business models

In addition to the initial conditions and the gradual policy shifts, the growth of the two sectors can be explained by specific industry logics such as the nature of the market and industry structure (including the protection of MUL for a decade before it was open to other multinationals and low entry barriers to the software

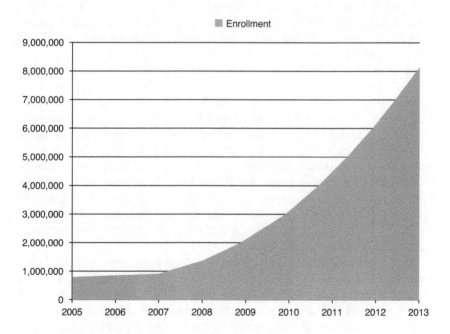

Figure 6.2 Undergraduate engineering enrollment (trends and forecasts)

Source: Rafiq Dossani 2010, "Understanding the Quality and Expansion of Higher Education in India, Guest Lecture," Asia Research Centre, Copenhagen Business School, September 2, 2010

industry). There is a third element and that is the dominant business model adopted by the industries. Both the auto and IT industry adopted flexible business models. The auto industry adopted a governance system that introduced production flexibility as practiced by the Japanese and adapted to Indian conditions, while the IT industry learned to quickly move from one kind of service delivery model to another, illustrating the flexible organizational capabilities and ability to ramp up production quickly.

In the auto industry Japanese industrial practices were imported by Suzuki and assimilated and diffused by MUL and its suppliers (see D'Costa 1998, 2004). The model can be described broadly as a new kind of "industrial governance," as the larger institutional context of the state–business relationship was under reconfiguration (D'Costa 2005: 105–141). Inheriting an industry that was technologically behind, the auto industry under MUL broke ranks with established practices. First, it introduced new capital equipment (technology) and a new fuel-efficient product.[12] Second, Suzuki introduced Japanese industry practices to secure quality components and ensure uninterrupted flow of production. This consisted of both inter-firm and intra-firm relations (D'Costa 2005: 109). Joint ventures between MUL and its suppliers were one such relationship. Several components producers also created joint ventures with Japanese suppliers, who were also suppliers to Suzuki in Japan. Many of these supplier units were co-located in the factory premises of MUL. In fact, MUL had a captive power source using in-house, large, oil-based generators to produce electricty, which it also shared with its suppliers (Gulyani 2001). Subcontracting arrangements, and a simpler version of the just-in-time system, were also adopted, with many suppliers co-locating in the vicinity of MUL's factories.

The intra-firm relationships included instituting teamwork, quality circles, multi-skilling routines, flexible machinery, and most importantly enterprise unions similar to Japan but different from the Indian practice of party-affiliated trade unions. The implication of the latter was to have a pliant workforce so as to not disrupt production in a high-growth economy. However, recently industrial strife has plagued the auto industry, especially those plants that are under multinational management (D'Costa 2011b). In several instances, employers have refused to recognize unions that do not have the blessing of the management. It is unclear whether this is a preemptive strike to prevent loss of managerial control over high-growth production or if there has been an intensification of work because of growing demand, with workers thus increasing resistance to managerial expectations.

The IT business model initially began with onsite services, whereby Indian professionals provided services at the clients' site, mostly in the USA. This was pejoratively known as "body shopping," since Indian firms were essentially brokers and recruiting agencies for technical staff. They billed projects on the simple formula of time spent on project by the number of people deployed, plus profits. The projects were simple mainframe computer-based and entailed low-value coding, debugging, maintenance, testing, and implementation. The profits for Indian firms could be linearly expanded if they could deploy more workers.

However, their costs were high, since Indian workers in the USA had to be paid close to US wage rates, as cost of living was high relative to India. With learning by doing and increasing "trust" by US clients, more software services were developed offshore in India. Now the projects handled are complex, large-scale enterprise solutions involving greater design, coordination, and team management. This is highly profitable since costs were much lower in India and the services provided were increasingly higher on the value chain with new domains of expertise. Consequently some of the bigger Indian firms also acquired foreign firms either to secure know-how or access markets. The USA is India's largest market for software services, absorbing over 60 percent of India's exports. India is the main supplier of technical professionals, capturing a third of all employer-sponsored H1B visas granted by the US government and with a high share of them joining the IT industry. India is also a major source of students for US universities, who later as expatriate professionals work for US firms. Some of these expatriates also return to launch IT businesses in India, thus reinforcing the industrial expansion trajectory.

The new business model also led to greater cooperation between firms and the state. MUL as a public sector unit already had a civil servant as one of two top corporate positions. Auto firms were members of either the Society for Indian Automotive Manufacturers (SIAM, previously AIAM) or the Automotive Components Manufacturers Association (ACMA). Both these bodies tried to influence policy pertaining to the auto industry. They were both linked to the Confederation of Indian Industry, the most powerful industry lobby representing newer engineering/entrepreneurial firms. In the IT industry, the influential National Association of Software and Services Companies (NASSCOM) enjoyed a privileged position. Not only was it a highly visible and proactive association in which more than 90 percent of Indian IT companies were members, it also tapped Indian emigrants in the USA to contribute to the flourishing offshoring business. The state–business relationship was also fundamentally restructured—from one of arms-length protection and regulation to one of collaborative partnership, from disdain of Indian emigrants to genuine acceptance of their role in the economic transformation process. The sheer visibility of the sector with techno-entrepreneurs, export success, and global links to centers of innovation such as Silicon Valley has lent a special political appeal and economic clout compared to other sectors. The IT sector was also governed by the presence of a non-unionized workforce on the premise that the IT worker is a well-paid professional, where the usual manufacturing or low-wage employment-related pay and working conditions issues do not arise. Industrial relations of this kind have been a necessary condition for the export success of the industry, where quality of work and rigid delivery times are crucial to capturing international markets. Millions of lines of code used for writing software on strict deadlines or fielding hundreds of international telephone inquiries for services demand a pliant workforce.

Based on these advantages, the Indian software industry has also been able to adjust quickly to changing global market conditions. For example, it began initially with on-site services by sending technical professionals to the USA on a short- and medium-term basis. The projects were simple, mainframe issues. However, profits were not high for the value of services provided, since costs of supporting professionals in the USA was much higher than in India. With experience and learning and increasing trust by foreign clients, the Indian software industry was able to work offshore in India, allowing for much better coordination and cost control, and thus significantly higher profit margins that were based less on the number of person-months dedicated to a project. The projects handled are also more complex than the typical on-site projects. Furthermore, the initial growth momentum reinforced expansion in education, mostly in the private sector, and significantly in southern states, where the education trajectory was already on a different track compared to other states. There is considerable effort by the Indian IT industry to move into more complex services, including engineering services and cloud computing on a big scale.

5   Conclusion: revisiting sectoral growth

This chapter illustrates the evolution of two high-growth sectors, auto and IT, in India. These were sectors for which India was not known, let alone being globally visible. The details of sector developments suggest that there are industry-specific factors at play, which might not allow easy generalization. However, in both sectors we observe gradual institutional change, that is, a re-specification of state–business relationships and market-based responses to new opportunities. The industry-specific factors in a changing institutional context have entailed: (i) initial conditions that place the industry in a particular trajectory once there is a reconfiguration of institutions around economic activities; (ii) industry logics that push or reinforce the industry on an expansionary path; and (iii) the adoption of flexible business models that can respond appropriately to the changed institutional environment.

The experiences of the two sectors also reveal that, contrary to other countries, India has continued to grow despite the global economic downturn. There were institutional buffers that could keep the industries going. Aside from the macro-economic conservatism, Indian industries were gradually opened up to the world economy. In other words, even when deregulation and liberalization radically altered the policy environment, the state continued to limit market entry in the auto industry and support the export-driven IT sector with infrastructure. While this does not reflect a coherent industrial policy, the necessity of giving space for industrial adjustment in a volatile situation is critical. It also suggests that simply paying heed to neoliberal dictates of budget cuts and a hands-off policy is inadequate for ensuring industrial dynamism; instead, more purposive forms of intervention for industrial growth and employment, though not sufficient, are necessary for social adjustment and stability.

At the industry level, the role of the state has been important. For example, the Indian state created a public sector–multinational joint venture. MUL enjoyed economic and technological advantages as the first comer in a newly organized industry and could build up scale economies. The government extended support through ad hoc but favorable policies. The company, through its Japanese partner, transformed the auto industry with the formation of a supplier sector, an outcome that was socially and economically desired. Local content rules forced the industry to develop local parts production capability. In one sense this case illustrates successful industrial policy. While the nationalization of Maruti was politically an expedient one, in the absence of clientelism and the presence of an expanding middle-class market, the company reaped substantial economic benefits. The Indian IT industry is a perfect example of the unwitting application of an earlier policy regime's need for technical professionals to a changed context, creating domestic capability in the hardware sectors. While rapid growth was predicated on IT industry demand in the USA, once the process was set in motion, Indian institutions, both public and private, got onto the bandwagon, reinforcing growth through investments in education and infrastructure. Once the industry was on a sustainable trajectory, the role of the state became promotional, pushing the sector onto an even higher growth path.

When growth is disaggregated at the sectoral levels, the role of the state can appear purposeful and effective, although the original rationale could be narrowly political. However, once the ball starts rolling, the relevant sectoral constituencies—the firms, industry associations, and sector-specific ministries—all try to work together to maximize their payoffs. As I have mentioned earlier, related sectors such as education also move in tandem. The development of the IT sector spurred institutional change, making state action more a product of industry growth than the other way round.

The question of clientelism is also an important one since it is rampant in India, although whether it was stronger during the earlier ISI regime is debatable. However, and consistent with the notion of effective states, the dominant public sector auto firm, unlike most other public sector firms, was given autonomy in the day-to-day operation of the company. An ambitious Japanese partner clearly set the terms as well. Clientelism was also absent within the company, leading to a more entrepreneurial management, including that of particular Indian civil servants (Bhargava 2010). In fact there are other, albeit few, examples, such as the metro system in Delhi, of capable bureaucrats successfully managing large industrial projects.

Market forces became important much later when the initial conditions, with some incremental reforms, had already prepared the groundwork for expansion. Both sectors adopted governance structures that allowed flexible adjustments to changing (mostly expansionary) markets and thereby gain from scale economies. Hence, it is no surprise that leading Indian firms from both auto and IT are acquiring foreign firms as a strategy for further expansion by way of accessing markets and technology. Expansionary growth has been possible in part because of Suzuki Motors, which diffused best industry practices and nurtured many supplier firms. Other firms experienced lower barriers to entry when the auto industry was liberalized in 1993. This might be termed the latecomer advantage.

The IT industry was already subject to US demand conditions. Buoyed by labor arbitrage, pricing was a not a significant issue. Labor was cheap and could be temporally exported in large numbers. However, what was critical was getting out of body shopping and fostering the offshore model with scale economies. Large firms were able to profit from their size. Learning by doing was critical to the sector's growth, and, once the state saw the potential of the software industry, it played a substantial role in providing IT infrastructure as well as fiscal incentives to exporters. The industry also took its own initiatives through self-regulation and representation (represented by NASSCOM, the software sector lobby) and ensured diffusing and adopting best (US practices). The industry also pushed the state to market and brand India, invest in technical institutions and infrastructure, and seek bilateral agreements on the mobility of Indian technical professionals in key IT markets. Here, too, many Indian firms have made strategic foreign acquisitions based on this learning. The software sector has been subject to a state–business partnership, with the latter calling the shots.[13]

Finally, both of these sectors can be clubbed as middle-class consumption and (for IT) production sectors. I would argue that this was a critical initial condition, which was instigated by state-led development under the ISI regime. Social differentiation in India has been the driver of the consuming and the educated classes, which are very much a post-independence development.[14] This differentiation has been reinforced further with high growth. The disjunction between what was possible to consume and what was available in India, combined with the demonstration effects let loose by petro dollars and the miracle economies of East Asia, pushed the levers of reforms, albeit haltingly. The ISI regime had created its own systemic interests and it would take at least a dozen years of expediency, experimentation, and tinkering by the state and business to alter their relationship in favor of business. It is in the 1980s (not the 1990s) that we witness the preconditions of industrial dynamism in the two sectors. It is clear that industry growth is not simply a matter of sound macroeconomic policy; rather, it is a net result of initial conditions, policies—both purposive and ad hoc—and industry logics that reinforce the trajectory. Where macroeconomic growth can help us understand industry dynamics is when such growth favors certain economic and social classes, as industrial development in India has done for the middle classes, providing consumer goods, services, and employment. The IT industry is a good example of how a diaspora and returned professionals reinforce the virtuous cycle of local education and growth via foreign education and expatriate entrepreneurs. An evolutionary perspective has helped in disaggregating these multifarious dimensions.

Acknowledgement

I would like to thank Janette Rawlings for her editorial support.

Notes

1   These have come to haunt India in more recent years, with the Indian Rupee taking a battering due to widening current account deficits brought on by massive imports; but this does not affect the main thrust of this paper on Indian industries.

2   See Sen (2010) for details of the debate as to whether the state or market (institutional change) led to high growth.
3   It must be stressed that between 1995 and 2002 the higher-income-earning households experienced much faster growth than lower-income households (National Council for Applied Economics Research and Business Standard 2004).
4   Several proposals, including one between TELCO and Honda, were ultimately rejected by the government of India on grounds of outflows of foreign exchange. The details of why Suzuki was selected and others rejected can be found in D'Costa (2005).
5   The government in the 1970s had established Scooters India Ltd. to capture the lucrative two-wheeler market. However, it failed miserably because of industrial strife and managerial and technological incompetence (Nayar 1992).
6   www.bangaloreit.com/html accessed on 04/15/2005.
7   http://en.wikipedia.org/wiki/Visvesvaraya_Technological_University accessed on 08/03/2011.
8   www.educationinfoindia.com/streamwisecolleges/others accessed on 04/18/2005.
9   Nayar (1992) illustrates the difficulty in privatizing state-run firms such as Scooters India Ltd. as part of the overall liberalization package. Even for private firms, labor retrenchment was fraught with difficulty, as in the case of Standard Motors.
10 It has been reported that Hyundai from its Indian operations has been exporting some cars and components back to South Korea.
11 Yogi 2014.
12 The Maruti 800 was a small car, often compared to a tin box. The model when it was introduced was on its way to being discontinued in Japan (Suzuki Plant Visit, Hammamatsu, December 1991).
13 In all of this, IT labor is in the background with no real political voice, a feature that is also true in the case of the auto industry, except for selective forms of labor resistance in particular auto locations (D'Costa 2011b).
14 The post-1991 development is also one of massive wealth and income inequality, which is also a middle-class phenomenon in India.

References

Amsden, A.H. (1989) *Asia's Next Giant: South Korea and Late Industrialization* (New York: Oxford University Press).

Bhagwati, J. (1993) *India in Transition: Freeing the Economy* (Oxford: Clarendon Press).

Bhargava, R.C. (2010) (with Seetha), *The Maruti Story: How a Public Sector Company Put India on Wheels* (Noida: Collins Business).

D'Costa, A.P. (2014) "Compressed Capitalism and the Challenges for Inclusive Development in India," in Dutta, D. (ed.) *Inclusive Growth and Development in Two Emerging Economic Giants of China and India in 21st Century* (Singapore: World Scientific and Imperial College Press, pp. 161–189).

D'Costa, A.P. (2011a) "Geography, Uneven Development and Distributive Justice: The Political Economy of IT Growth in India," *Cambridge Journal of Regions, Economy and Society*, 1–15, doi:10.1093/cjres/rsr003.

D'Costa, A.P. (2011b) "Globalization, Crisis and Industrial Relations in the Indian Auto Industry," *International Journal of Automotive Technology and Policy*, 11 (2), 114–136.

D'Costa, A.P. (2009) "Extensive Growth and Innovation Challenges in Bangalore, India" in Parayil, G. and D'Costa, A.P. (eds) *The New Asian Innovation Dynamics: China and India in Perspective* (Basingstoke: Palgrave Macmillan, pp. 79–109).

D'Costa, A.P. (2005) *The Long March to Capitalism: Embourgeoisment, Internationalization, and Industrial Transformation in India* (Basingstoke: Palgrave Macmillan).

D'Costa, A.P. (2004) "Flexible Institutions for Mass Production Goals: Economic Governance in the Indian Automotive Industry," *Industrial and Corporate Change*, 13 (2), 335–367.

D'Costa, A.P. (2000) "Capitalist Maturity and Corporate Responses to Liberalization: The Steel, Auto, and Software Sectors in India," special issue on Corporate Capitalism, *Contemporary South Asia*, 9 (2), 141–163.

D'Costa, A.P. (1998) "An Alternative Model of Development? Cooperation and Flexible Industrial Practices in India," *Journal of International Development*, 10 (3), 301–321.

D'Costa, A.P. (1995) "The Long March to Capitalism: India's Resistance to and Reintegration with the World Economy," *Contemporary South Asia*, 4 (3), 257–287.

Doner, R.F. (2009) *The Politics of Uneven Development: Thailand's Economic Growth in Comparative Perspective* (Cambridge: Cambridge University Press).

Encarnation, D. J. (1989) *Dislodging Multinationals: India's Strategy in Comparative Perspective* (Ithaca, NY: Cornell University Press).

Evans, P.B. (1995) *Embedded Autonomy: States and Industrial Transformation* (Princeton, NJ: Princeton University Press).

Grieco, J. (1984) *Between Dependency and Autonomy: India's Experience with the International Computer Industry* (Berkeley, CA: University of California Press).

Gulyani, S. (2001) *Innovating with Infrastructure: The Automobile Industry in India* (New York, Palgrave Macmillan).

Heeks, R. (1996) *India's Software Industry: State Policy, Liberalisation and Industrial Development* (Thousand Oaks, CA: Sage).

Heitzman, J. (2004) *Network City: Planning the Information Society in Bangalore* (New Delhi: Oxford University Press).

Indiastop.com (1998) "India's Consumer Markets: Identifying a Plausible Market Size of Products," www.Indiastop.com, accessed 02/14/2002.

Kathuria, S. (1996) *Competing through Technology and Manufacturing: A Study of the Indian Commercial Vehicles Industry* (Delhi: Oxford University Press).

Kathuria, S. (1990) *The Indian Automotive Industry: Recent Changes and Impact of Government Policy* (New Delhi: A Study Prepared for the World Bank).

Kohli, A. (2007) *State-directed Development: Political Power and Industrialization in the Global Periphery* (New York: Cambridge University Press).

Kohli, A. (1992) *Democracy and Discontent: India's Growing Crisis of Governability* (Cambridge, Cambridge University Press).

NASSCOM (2014) "Industry Rankings," www.nasscom.in/industry-ranking, for 2007–08 and 2012–13, accessed 05/06/2014.

National Council for Applied Economics Research and Business Standard (2004) *The Great Indian Middle Class: Results from the NCAER Market Information Survey of Households* (New Delhi: NCAER).

Nayar, B.R. (1992) "The Public Sector," in Gordon, L.A. and Oldenburg, P. (eds) *India Briefing 1992* (Boulder, CO: Westview Press, pp. 71–101).

North, D.C. (2010) *Understanding the Process of Economic Change* (North Princeton, NJ: Princeton University Press).

Okada, A. (2004) "Bangalore's Software Cluster: Building Competitiveness through the Local Labor Market Dynamics," in Kuchiki, A. and Tsuji, M (eds) *Industrial Clusters in Asia: Analyses of their Competition and Cooperation* (Tokyo: Institute of Developing Economies and Japan External Trade Organization, pp. 276–314).

Pinglé, V. (1999) *Rethinking the Developmental State: India's Industry in Comparative Perspective* (New York: St. Martin's Press).

Rajaraman, V. (2012) *The History of Computing in India, 1955–2010* (Bangalore: Supercomputing Education and Research Centre, Indian Institute of Science).

Rediff.com (2011) "Top 20 IT Companies in India," August 3, www.rediff.com/business/slide-show/slide-show-1-top-20-it-companies-in-india/20110803.htm, accessed 08/03/2011.

Reserve Bank of India and Ministry of Human Resource Development, Govt. of India (ON117) in Indiastat.com, www.indiastat.com/default.aspx, accessed 03/15/2010. Note: Indiastat.com is a database.

Rueschemeyer, D. (2009) *Usable Theory: Analytic Tools for Social and Political Research* (Princeton, NJ: Princeton University Press).

Sen, K. (2010) "New Interpretations of India's Economic Growth in the Twentieth Century," in D'Costa, A.P. (ed.) *A New India?: Critical Reflections in the Long Twentieth Century* (London: Anthem Press).

Sridharan, E. (1996) *The Political Economy of Industrial Promotion: India, Brazilian, and Korean Electronics in Comparative Perspective, 1969–1994* (Westport, CT: Praeger Publishers).

Yogi, V. (2014) "Performance Review: Indian Auto Component Industry in 2013–14," July 17, http://auto.ndtv.com/news/performance-review-indian-auto-component-industry-in-2013-14-589644, accessed 05/11/2015.

7　A system abandoned

Twenty years of management, corporate governance and labour market reforms in Japan

Holger Bungsche

1　Introduction

In the 1980s, Japanese-style management was taught in all leading business schools in the USA, the UK and elsewhere in the Western world; books on the subject filled long shelves in bookshops and libraries, and company managers from all over the world flew to Tokyo and Osaka to see and learn Japanese-style management and lean production organization on site.[1] And also for people with no specific interest in management, the success of Japanese companies was visible, as the Honda Cubs, the Sony Walkmans, the Panasonic video recorders and the Toyota Corollas conquered the product markets worldwide. It seemed obvious that the Japanese company and management system was somehow superior to Western ways of management. Life-long employment, seniority-based promotion and remuneration, company unions, cross-shareholding between companies and financial institutions, as well as internal financing by one main bank – the main pillars of the Japanese system – made the companies largely independent of outside labour markets and in particular freed them from the pressure of the financial market to deliver short-term results to satisfy shareholders and stockmarket brokers. This, it was said, enabled Japanese companies to first educate and then bring the best talents into management positions, to strategically develop the business with a long-term perspective, focusing on increasing future market shares instead of quarterly profits.

Very much like in Germany, where in the 1950s under the minister for the economy Ludwig Erhard the social-market economy was established with the intention not only to increase the living standard of the population considerably, but also to make the economy as a whole more democratic, the Japanese company and management system evolved after the Second World War out of the endeavour to completely rebuild and modernize the country, both economically as well as politically. For both, the efforts of the whole population was needed, which of course also made it necessary that the economic gains were distributed equally. And like the German *Wirtschaftswunder* (economic miracle) in the 1950s and 1960s, Japan at the end of the 1950s entered into its high-growth era, laying the foundation which saw the country develop into the second largest economy in the world in the 1980s.

However, the Japanese model was not only economically, but also socially, very successful. Based on a very equal income distribution, overall living standards improved drastically – in regular surveys 90 per cent of the population in the 1970s/80s were perceived by themselves to belong to the middle-class – while poverty, homelessness and unemployment were almost unknown. Crime rates were the lowest in the world, just as literacy rates, percentages of the population finishing college or university education, and life expectancy ranked amongst the highest.

So, in short, Japan provided a viable and very attractive alternative to the Anglo-Saxon type of capitalism. This, however, changed in the 1990s after the bursting of an economic bubble that resulted from the overheating of the stock and real estate markets. In reaction to the bursting of the bubble economy, Japan moved considerably away from its own model, 'implementing international standards', as these were called. Specifically, with the restructuring of the financial sector, Japanese company regulations were modified in a way that completely altered the shareholder structure of Japanese listed corporations. Not least because of the changing shareholder structure, company management, especially with regard to evaluation, remuneration and promotion of employees, changed quite drastically. The reforms of the company law and the implementation of new corporate governance standards were accompanied by a far-reaching liberalization of the labour market, particularly with respect to non-regular forms of labour. By these reforms the traditional Japanese management and company system was basically undermined. The reforms were politically justified by the slogan, 'economic recovery by easing regulations', while the economic explanation, especially from the companies, was that

> the Japanese system was good in times of growth, as there was a large pie that could be distributed; now in times of mega-competition, however, there is no pie that could be distributed, so one had to switch to international (Anglo-Saxon) standards.

In the following we will first briefly give an overview of Japan's economic development, employment and industry structure. Based on this we will discuss the influence of the bursting of the bubble economy on the Japanese company system, in particular with respect to (1) the restructuring of the financial sector, new company legislation and attempts to implement a corporate governance code, (2) management changes in evaluation, remuneration and promotion, and finally (3) the deregulation of the labour market. From today's perspective in Japan, the financial crisis of 2008 was just a singular, short-term incident that did not really alter the overall course towards 'liberalization'. However, there are signs that the financial crisis might have contributed at least at the company level to some kind of reassessment of management changes that had been implemented in the previous two decades. Whether, and to what extent, the 2008 financial crisis has led to a significant reevaluation is addressed in a brief outlook for the future at the end of this chapter.

2   Japan's economic development: GDP growth, industry and employment structure

Before we turn our attention to the paradigmatic shift in Japan's management and company system, we first look at the overall economic development of Japan since the Second World War. Based on Japan's gross domestic product (GDP) growth rates, we can clearly distinguish between three stages, as shown in Figure 7.1. First, the 'catching up' phase, with average growth rates above 9 per cent from the mid-1950s until the first Oil Crisis of 1973. Second, a maturing period, from the mid-1970s until 1990, with growth rates at about 4.2 per cent per year, during which Japanese companies pushed their globalization strategies by investing and increasing production capacities abroad. And finally, the third stage running from 1990 and the bursting of the bubble economy, during which period domestic economic growth slowed down to only 0.9 per cent on average per year.[2]

With the bursting of the bubble, Japan's economy entered the so-called 'lost decade'.[3] In the years between 1995 and 2005 economic growth was almost entirely export driven; domestic consumption and consumer prices stagnated or declined, and Japan had to face several periods of deflation.[4] Besides the bursting of the bubble economy, there are of course other reasons for the slowdown of GDP growth as well. First, Japan's economy has grown into a mature

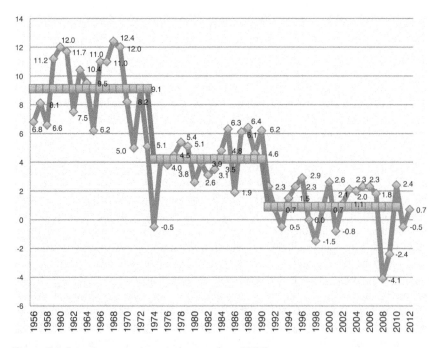

Figure 7.1 Japan's economic growth rates since 1956[5]

Source: Cabinet Office, Government of Japan

economy; second, Japan's demography is quickly changing;[6] and third, Japan is increasingly facing competition from emerging economies, especially in neighbouring Asian countries.

Regarding the economic impact of the financial crisis of 2008, we see a decline of more than 4 per cent of GDP in that financial year.[7] Differently to the USA and the EU, however, Japan's financial sector was almost unaffected by the US subprime crisis. In Japan the financial crisis hit the real merchandise economy, especially the exporting industries. Maybe because of this the crisis was not perceived as a systemic crisis challenging Japan's economic fundamentals, as it was – at least temporarily – in other countries where the financial sector was affected.

Looking now at industry structure and employment, we see first that Japan still has a very manufacturing-based economy, very similar, for instance, to Germany. In 2013 agriculture contributed 1.2 per cent, the production industries 24.5 per cent and services 74.3 per cent to Japan's GDP (CIA World Factbook). Also, the economic structure of Japan, which is characterized mainly by small and medium-sized companies, is very similar to Germany's structure, where the so-called *Mittelstand* also forms the backbone of the economy.

Regarding the population in work, we see that the number of people in work increased until the second half of the 1990s. After the bursting of the bubble economy, employment decreased slightly but has remained quite stable ever since at about 62 million persons. In fact, neither the burst bubble economy in the mid-1990s nor the financial crisis of 2008 had a significant impact on overall employment, as Figure 7.2 also illustrates.[8]

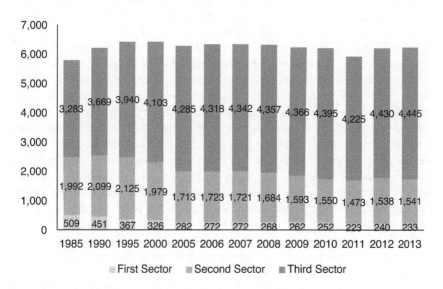

Figure 7.2 Development of sector employment in Japan

Source: Ministry of Internal Affairs and Communication, Statistics Bureau

Looking at sector-specific employment, we see that manufacturing as well as wholesale and retail trade are providing the most jobs to people, both about 11 million. In these industries, especially manufacturing, small and medium-sized companies are providing by far the most jobs. Roughly two-thirds of all employees in Japan are working for small and medium-sized companies. Since 2002 the share of large-scale enterprises is increasing, but employment in medium-sized enterprises especially is very stable.[9]

Finally, when looking at the unemployment statistics, we see that between 1980 and 1995 Japan's full unemployment rate was constantly between 2 per cent and 3 per cent. After the bubble burst, the rate increased up until 2003, to 5.3 per cent. In the following five years, and until the world financial crisis, unemployment fell slightly to below 4 per cent, but increased again to 5.1 per cent in 2010. From May 2010, however, it decreased for 60 consecutive months and is currently 3.3 per cent (May 2015).[10]

Based on these macro-economic, industry and employment data, we will now turn to the paradigmatic shift away from the Japanese traditional economic model that occurred after the bursting of the bubble economy.

3   The bursting of the bubble economy and its influence on the Japanese company and management system

The bubble economy has been widely perceived as a radical incident, a turning point in the economic development of Japan. This it definitely was. However, we should not so much concentrate on this singular incident, but on the broader long-term development. Based on the above-presented macro-economic data we can state that Japan's economy at the end of the 1980s was mature and fully developed. And, with that, it was no longer in the phase of catching up with other economies, but more and more subject to outside competition itself, especially from neighbouring Asian emerging economies. So Japan had to compete with other production locations; in particular it became increasingly confronted with a shift of markets and production locations away from Japan to neighbouring Asian countries, first and foremost to China, but increasingly also to the ASEAN member states.[11] At the same time, while domestic wage and production costs continuously increased during the growth era between the 1950s and 1980s, since the 1990s domestic consumption has been stagnating. Finally, on the world markets Japan's companies have, since the 1990s, been increasingly confronted with new competitors – especially from emerging countries like Korea or Taiwan that are challenging Japanese companies not only on price, but also more and more on quality.[12]

So, to put it simply, Japan and its companies faced in the 1990s, and for the first time, the challenge of how to stay competitive under the conditions of a quickly and profoundly changing economic environment. And, as we will see below, with pressure coming from business associations, companies, financial markets and other interest groups, the Japanese government prepared the institutional framework for the implementation of perceived global standards, which led to a paradigmatic shift in Japan's company and management system as well as in its labour market.

3.1   Restructuring of the financial sector, company law reforms and new standards for corporate governance

The paradigmatic shift that occurred after the bubble economy burst put an end to the 'traditional' Japanese management and company system that had evolved after the Second World War. From the late 1950s, until the bursting of the bubble, Japan's social-economic model could be described as, and what Ronald Dore called, 'stakeholder capitalism'. The main feature of this 'stakeholder capitalism' was that 'quasi employee community firms' aimed at increasing production and market share in order to redistribute the gains to their employees and with this to society, which thus became a 'stakeholder society'. This system was primarily based on four pillars: (1) lifetime employment, (2) seniority-based salaries, (3) company unions and (4) the main-bank system as well cross-shareholding.[13]

Although lifelong employment in fact was never formally guaranteed and although it roughly applied to only 30 per cent to 40 per cent of all company employees, it provided on the one hand employment security, stable career development and, with the number of working years progressively increasing, rising employee income.[14] On the other hand, the internal labour market resulting from the lifetime and seniority-based employment system, together with the main-bank system, where company-affiliated banks held the shares of companies and, vice-versa, the companies held shares in these banks, made the companies largely independent both from financial markets and outside labour markets. This is what gave the companies the freedom to concentrate on long-term company development and growth instead of concentrating on short-term returns on capital to raise shareholder dividends.[15]

However, in the post-bubble period, the traditional Japanese company system began to change fundamentally due to new company legislation, a complete restructuring of the banking sector and, last but not least, broad changes in companies' employee evaluation and remuneration systems.

The emergence of the bubble economy and the economic damage caused in its aftermath was – for whatever reason – solely blamed on this traditional Japanese company and management system. One main argument was that the Japanese company system was not compatible with international corporate governance standards. More specifically, because of the close interconnection between companies and financial institutions, it was impossible to apply effective control mechanisms that would have been able to adequately evaluate the companies' exposure to external risk factors. There was, so the argument went, simply no control exerted, either at the company level or the company group level (*keiretsu*). Numerous scandals after the bubble burst revealed that there was indeed widespread corporate misbehaviour, insufficient risk evaluation and control mechanisms as well as a general lack of transparency and accountability at many companies, especially in the financial sector. Beginning in 1997, this led to an intensive discussion about corporate governance reforms in Japan. It was argued that Japanese companies had to apply international corporate governance standards, since – as the then new 'Japan Corporate Governance Forum' stated – decent corporate governance and compliance with international standards are a crucial aspect for company success in the age of globalization (Buchanan and Deakin 2009: 31–34).

In particular there were two waves of reforms and amendments of company-related legislation and corporate governance regulations in Japan. The first wave, resulting from the discussion started after 1997, led to the amendment of the Commercial Code in 2002, which allowed companies to introduce an alternative structure of the board of directors, the so-called 'companies with committees'. Companies that changed their board structure usually reduced the number of board members considerably and appointed a larger number of independent external directors. At the same time the power of the board of auditors, which has the task of monitoring management, was strengthened. While the number of companies that formally introduced the 'companies with committees' board structure was still limited, many more companies altered their board structure by introducing 'company executive officers' and separating executive management functions from supervisory functions. Amongst these companies were Toyota and Canon (Buchanan and Deakin 2009: 34–39).

In July 2005, the proposals of the Corporate Governance Forum were taken up, and various different regulations and laws related to company governance were finally put together in one piece of legislation, the Japanese Companies Act, which was enacted in 2006.

The second wave set off in autumn 2011, when the first amendments of the new Companies Act were proposed. The discussion about the amendments gained momentum when in 2012 the LDP (Liberal Democratic Party of Japan) again took over government and Prime Minster Abe perceived corporate governance reforms to be an appropriate tool for pushing a wider spread of structural reforms in Japan, which he promulgated as the third arrow of his economic revival strategy for Japan, called 'Abenomics'.[16] Corporate governance reforms were perceived to be key for increasing the productivity and profitability of Japanese companies, in order to stimulate overall economic growth leading to sustainable economic recovery after the so-called '20 lost years'. As a result of these reform efforts, the Company Law was amended in June 2014. Based on a 'comply-or-explain' approach, the new company law proposes the appointment of at least one external director, strengthens internal control mechanisms and sets the highest priority to increasing productivity and earnings by improved corporate governance. Simultaneously, a new stock index, the JPX-Nikkei 400, was introduced, which is comprised of companies that have implemented new corporate governance standards and are recording superior economic performance (return on equity and profits). Finally, company and investor relations are also to be newly regulated. The basis for this is a voluntary adoption of a 'Stewardship Code' similar to in the UK, and for the first time a comprehensive Corporate Governance Code has been drafted (Freshfields Bruckhaus Deringer 2015: 1, 3).

The second point that especially affected the traditional cross-shareholding and main-bank system was the restructuring of the financial sector in Japan after 1997. After the bubble, Japan was confronted with the problem of non-performing loans caused mainly by the so-called 'Jusen banks', which were non-bank financial institutions specialized in real estate speculation. Since commercial banks were not allowed to engage in real estate business, the 'Jusen-banks' took over the real estate business; however, the financial resources for doing so came from closely affiliated regular banks. It took several years until the 'Jusen' problem was resolved, basically by bailing out the banks using budgetary resources.[17] According to the Financial Service Agency (FSA), accumulated losses due to the disposal of non-performing loans between 1992 and 2004 amounted to 96,420 billion yen. At the same time, in the second half of the 1990s the bank deposit insurance system was strengthened and procedures improved to deal with and speed up bankruptcies.

Finally in preparation for liberalizing the financial sector, a new dual system for supervising the banking sector was implemented. Based on the Japanese Banking Act, the Financial Service Agency (FSA) was established as an independent institution, replacing the Ministry of Finance as the top supervising body for the banking industry.[18] Simultaneously with this legally binding supervision system a second 'voluntary' system of supervision was established. Based on the regulations of the Bank of Japan Act and bilateral agreements by all banks that have current accounts with the Bank of Japan (BOJ), which in fact include almost all domestic as well as foreign banks, the BOJ was entitled as well to conduct on-site examinations of financial institutions in order to maintain a safe and sound financial system, including examinations of banks' overseas branches.

In 1998 the financial sector was finally liberalized. In a first step, the Foreign Exchange Law was reformed and the until then required procedures of prior application and checks for international transactions were abandoned. In a second move, all inter-business barriers between banking, insurance, securities and asset management businesses were removed. This liberalization led to the so-called Japanese 'Big Bang' of 1998 and to a complete restructuring of the financial sector from the year 2000 on. Due to the liberalization, the number of country-wide operating city-banks more than halved from 13 companies in 1990 to only 6 today, and the number of smaller regional banks decreased from 66 institutions to 44 (Japanese Bankers Association).[19] The effect of this reform and restructuring process after the bursting of the bubble economy, however, was that the Asian Financial Crisis of 1997/98 and the US subprime crisis ten years later left the Japanese financial sector almost entirely unaffected.

Looking at the effect on company management and the company system in Japan, the new company legislation and corporate governance reforms on the one hand, as well as restructuring of the banking industry on the other, resulted in a complete new shareholder structure. Cross-shareholding between companies or between companies and financial institutions decreased dramatically, while the percentage especially of institutional and foreign investors grew tremendously. According to a study published by the Research Institute of Economy, Trade, and

Industry (RIETI), within not even two decades the share of institutional share-holders more than doubled from 23 per cent in 1996 to 48 per cent in 2013 (Miyajima 2014: 2). As Figure 7.3 shows, this change occurred at around the year 2000 as the share of outside shareholders, which means the share in the hands of investment and pension trusts, and individual and foreign shareholders, began to exceed the share held by insiders, mainly city and regional banks, life and non-life insurance companies, business corporations, etc. In particular, the banks and insurance companies, which in 1985 held more than 37 per cent of the shares of companies listed on the Tokyo Stock Exchange, reduced or had to reduce their capital drastically. In 2014 the percentage held by banks and insurance companies was just 8.7 per cent. And also business corporations, which in the 1980s also held more than 30 per cent of shares, sold more than a third of their investments. This, as indicated above, brought an end to cross-shareholding, which was once perceived as a decisive component of the Japanese management and company system, because the stable investor relations that it provided allowed the compa-nies to pursue a long-term company development strategy without the pressure coming from shareholders that demanded short-term profits.

Together with the shareholder structure, the distribution of company profits also changed completely. Traditionally, Japanese companies' preferences with respect to redistribution of profits were first of all to secure necessary investments in the modernization of production equipment, etc. for expanding market share in the future; second, to distribute a large share of the profits to the company's employees who created these profits; and finally, to give a smaller, but fair share

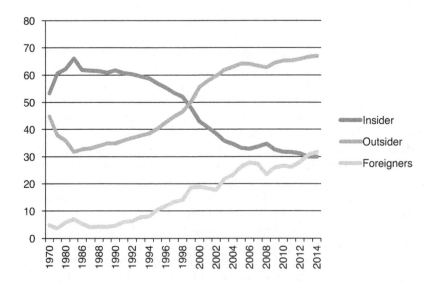

Figure 7.3 Changing company ownership

Source: Tokyo Stock Exchange Shareownership Survey 2014[20]

to the shareholders, who provided the money for running the business. With the change in ownership, however, after 1990 the pattern of distribution of profits between investors and company (employees) also began to shift from the company side to the investors' side, as the Table 7.1 shows.

Undoubtedly, the changing shareholder structure placed tremendous pressure on the companies to redistribute profits far more to the capital owners than before. This also becomes obvious when looking briefly at the discrepancy between the development of employee wages and shareholder dividends in the time period around the bursting of the bubble economy in the early 1990s and after the restructuring of the financial sector in the early 2000s (Table 7.2).

As the data indicate, there is a clear shift of income distribution observable, away from the labour side towards the capital side. This shift occurs around 1995, as the above-outlined reforms of company law and corporate governance and the restructuring efforts of the financial sector began to take shape. The paradigmatic change becomes even more evident when looking at the macro-economic development of Japan over the last two decades. Despite the often-cited phrase of the lost ten or 20 years, according to Organisation for Economic Co-operation and Development (OECD) data, productivity in Japan increased by about 2 per cent on average yearly after 1990, and GDP grew by roughly 12 per cent between 1995 and 2012. This means that the labour side did not profit from productivity gains or from the overall economic development.

Apart from the argument that Japan has to adapt to international standards, which was in general brought forward for justifying the reforms to company law, corporate governance and the financial sector, it was also, especially from the side of the Japanese companies, argued that the traditional Japanese management and company was appropriate in times of economic growth and in times where the task for Japan was to catch up with other developed economies. In these times there was

Table 7.1 Distribution of profits between investors and company

	Investors	*Company*
1980	10%	90%
1990	17%	83%
after 2000	27%	73%

Source: Dore 2006 and 2010

Table 7.2 Development of wages and dividends

	Wages	*Dividends*
1987–1992	+20%	+2%
2000-2006	-4%	+196%

Source: Dore 2006 and 2010

a steadily growing 'pie' that could be distributed amongst the various stakeholders. However, Japan is now a developed economy and Japanese companies have to survive in an environment of global mega-competition. In this environment the traditional Japanese management and company system is no longer suitable, since the 'pie' that can be distributed is not growing any more. Therefore, choices have to be made with respect to the distribution of the increasingly shrinking pie.

This now brings us to the issue of the changes implemented by the companies with respect to their internal evaluation and wage systems after the bursting of the bubble economy.

3.2 Changing company management: evaluation, eremuneration, and promotion

Although the bursting of the bubble economy was mainly a crisis of the financial sector that had only limited influence on the merchandise economy, many companies introduced new evaluation and remuneration standards that were no longer solely based on years of company attendance, but on personal achievement. Many younger employees and managers, who in fact did much of the work, but received less payment than their senior superiors or co-workers, welcomed this change. However, only a few really enjoyed a significant increase in their salaries. Most employees were not better off after the change of the evaluation and salary system in the companies.

We will concentrate on a few aspects. First we look at the effect on first-time salaries for university and high school graduates. Despite what has been elaborated above, there are still some components of the traditional Japanese system that have not changed considerably. One of these is the labour market for full-time, regular, permanent employment after graduation from university, which still is the only real free, external labour market, where the university graduates fight for the best jobs at the most prestigious companies. Once the graduates have taken their decision which company they join, the only way for advancing their careers in principle is within the company. Apart from some high-achievers, who will be approached by head-hunters, there is no external labour market for permanent positions.

Therefore the salaries for newly recruited employees are of outstanding importance, because the entrance salaries provide the basis for their future salary development.

From Table 7.3 one can see that between 1955 and 1995, first-time wages for university graduates increased 15 times, while prices only increased six times during the same time. Since, over the same time period, Japan's GDP only increased ten times, it is appropriate to state that until the mid-1990s income distribution favoured the labour side and that also income inequalities were relatively low.[21] This, however, as already stated before, began to change after the bursting of the bubble economy, when most listed companies shifted their evaluation and payment schemes at least partially from seniority- to achievement-based schemes, which can also be seen from the data in Table 7.3 for new company employees. The increase after 1995 is marginal. Of course, since prices also remained stable,

Table 7.3 First-time salaries for high school and university graduates[22]

	First salary of university graduates	First salary of high school graduates	Consumer price index
1955	12,907	11,489	297.4
1960	16,115	9,655	328.0
1965	24,102	17,023	443.2
1970	40,961	32,063	577.9
1975	91,272	74,713	985.3
1980	118,138	96,955	1364.6
1985	144,541	116,830	1578.8
1990	173,996	136,946	1709.8
1995	198,063	156,074	1829.5
2000	201,389	158,866	1842.8
2005	203,230	159,037	1785.2
2010	207,445	162,749	1767.3
2012	207,585	162,369	1739.8
2014	209,868	164,149	n.a.

Source: Salaries until 2000 taken from Bukka no bunkashi jiten (2008); salaries after 2000 taken from Keidanren (2015); price index taken from Bank of Japan

employees suffered no real loss in purchasing power, but they also did not participate in the GDP and productivity increases realized after 1995.

However, if nominal wages did not decline, or just temporarily declined, business-results-related bonus payments (paid twice a year) were cut quite drastically. So when we are looking at the development of wages after the bursting of the bubble economy we observe continuously decreasing real wages. In the years before the bursting of the bubble economy, bonuses usually amounted to 5 or 6 months' salaries, while after the mid-1990s they were reduced to often not even half of the former amount.[23]

As indicated above, many young employees and managers were very much in favour of changing the seniority-based wage system to a system much more based on achievement evaluation. However, with the shift from seniority- to achievement-based wage systems after 2000 the progression of salaries flattened. This affected especially the employees and middle-managers in their mid-life careers. Traditionally in Japan the progression of salaries was the fastest and the highest for employees between the age of 35 and 55. Usually, the employees of this age group have children, for whom they have to pay high tuition fees for school or university education, and they often have to pay back loans for their houses or apartments. As Table 7.4 clearly indicates, until the end of the 1990s wage progression for employees over 40 years old was still increasing, while the progression for younger employees was already beginning to flatten. After the turn of the millennium, wage progression across all age groups is considerably declining. And as can be seen for the age groups above 35 years, it was and is especially the employees in the middle of their careers who suffered the most significant losses in income due to the change of the evaluation system from seniority- to achievement-based evaluation.

Table 7.4 Development of salary progression of male production worker between 1989
and 2014 (index)

	1989	1994	1999	2004	2009	2014
20–24	100	100	100	100	100	100
25–29	123	122	120	115	115	112
30–34	151	148	148	133	130	126
35–39	175	171	172	148	143	138
40–44	197	189	190	160	157	152
45–49	212	205	203	169	164	161
50–54	209	210	211	177	171	169
55–59	181	189	198	174	173	166
60–64	145	146	147	127	125	115

Source: Ministry of Health, Labour and Welfare, Basic Statistical Survey on the Wage Structure;
several years

As a result of the bursting of the bubble economy and the above-mentioned changes in evaluation and remuneration, wage progression of ordinary employees in the companies flattened. This, together with nominal wages that after 1998 did not increase any more, and with considerably lower annual bonus payments, led to shrinking household incomes.

Since, in contrast to ordinary employee salaries, remuneration and bonuses of top managers and executives over the past decades has increased continuously, the gap between ordinary employees' and top management's income has widened considerably. Traditionally in Japan the gap between average ordinary employee salaries and top management remunerations had been very small, and of course Japan is still far away from a relation of 1 to 500, which is the discrepancy between ordinary employee and top management remuneration in the USA; but the gap is widening, and the traditional ratio of 1:10 or 1:15 is already a thing of the past. According to Dore, management bonus payments doubled from a ratio of 1:2.5 to 1:5 in relation to employee bonuses after the year 2000. Stock options as well as share buy-backs, which were legally forbidden until a few years ago, have also become quite widespread in Japanese companies as parts of management remuneration.[24]

Of course, one could state that Japan, like many other advanced economies, came under pressure for modernization due to increasing global competition, or to put it simply, due to globalization. That definitely is not wrong. However, Japan did not adjust its system to the changing global environment, but to a large extent simply abandoned the old system. This will become even more obvious when we look at the reforms and changes of the labour market.

3.3 Deregulation of the abour market

As early as the mid-1980s, first measures for deregulating and liberalizing Japan's labour market were introduced. The aim of these reforms was enhancing labour market flexibility. Not to be misunderstood, there have always been non-regular forms of work in Japan like, for instance, part-time work or tempo-rary work. And, as also stated before, lifelong employment was de facto only guaranteed to core employees of large corporations. Non-core employees, espe-cially female employees, provided the companies with the necessary flexibility to adjust the workforce to macro-economic changes or fluctuating demand in the export markets.

Therefore, various forms of non-regular work already existed before the revi-sion of labour legislation in the mid-1980s. However, these non-regular forms of employment were relatively limited in scale and never exceeded about 15 per cent of the workforce. Not all people who worked part time did so at their own discretion; however, in the case of part-time work at least, in the 1980s it was chosen to a much larger extent because it met the needs of the respective employee rather than because there was no other choice: female employees who wanted to earn some additional money for themselves or for the household budget and at the same time care for their children or look after an elderly family member; veteran employees, who wanted to continue work after official retire-ment and help with their knowledge and experience to improve management at a lower ranked company within a company group. Of course, no false romanti-cism is merited at this point: there were many people who had to continue working after retirement, as for example in the construction sector, because their retirement allowances were not sufficient; and there was also precarious work; and there was the so-called 3-k work (*kitsui* – hard; *kitanai* – dirty; *kiken* – dangerous) that was often done by day labourers or illegal foreigners who were collected in the early morning hours by the 'employer' at special locations in the suburbs of Osaka or Tokyo.[25]

However, with the labour market reforms, which will be presented below, and especially with the bursting of the bubble economy, Japan's labour market began to profoundly shift away from regular to non-regular forms of employment, and more and more people became dependent on non-regular work assignments for making a living. This means that, beginning in the 1990s, various forms of precarious work – that done by the working poor, or people needing two or more jobs to survive and other phenomena usually associated with developing coun-tries or neo-liberal Anglo-Saxon style capitalism – also took root in Japan's labour market.

Looking at employment data, we see that until the early 1990s non-regular employment accounted for just about 10 per cent to 15 per cent of all employ-ment. However, between 1990 and 2000 the proportion of non-regular employment increased to about 25 per cent. This means that many newly created jobs were just part-time jobs and that many companies, often regardless of their actual economic conditions, substituted jobs for regular employees with jobs for non-regular ones. This development even accelerated after the year 2000 as more and more professions were opened up for agency work. This trend continued almost uninterrupted by the world financial crisis and, with almost 38 per cent of the workforce working in non-permanent jobs in 2014, non-regular forms of work have increasingly become the rule rather than the exception, as indicated in the

Figure 7.4 showing development of regular and non-regular employment since the mid-1980s.

Before we turn our attention to the labour law reforms, we first have to explain the various non-standard forms of work that exist in Japan and how one distinguishes between the various categories. The most common form of non-regular work is part-time work. The worker is directly employed by the employer, both on fixed-term or open-ended contracts and is not working full time, but part time. Temporary work (in Japanese, *arubaito*, derived from the German word *Arbeit*) is often full-time work, but on relatively short fixed-term contracts. Contract and entrusted employees refers to employees who are usually working full time on relatively long fixed-term contracts; entrusted and contract employees are also each directly employed by an employer.[26] And finally, agency or dispatched workers are dispatched by private work agencies to client companies, where they work both full time or part time. Since the employer, however, is not the client company, but the work agency, this is a form of indirect employment. Usually contracts are fixed term, but there are also open-ended contracts.[27] However, there are many grey areas and in each of the above categories one finds decent, but also precarious, employment.

As can be seen from the Figure 7.5, until 2009 it was in the first instance (presumably) precarious forms of non-regular employment, especially agency and contract work, that increased disproportionately fast; but after 2009 part-time work and to a minor degree also *arubaito* (temporary work) began to rise again.

The crucial point that triggered this development was the opening of the labour market for private work agencies, which in a way institutionalized precarious work in Japan. This happened with the new legislation on dispatched workers of 1986 (Worker Dispatching Act) and its amendments of 1999 and 2004, respectively.

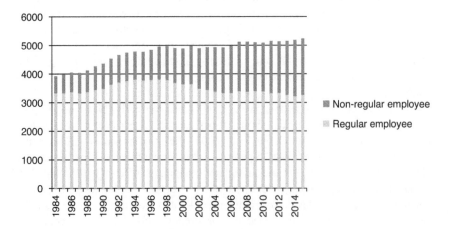

Figure 7.4 Development of regular and non-regular forms of employment

Source: Ministry Internal Affairs and Communication, Statistics Bureau

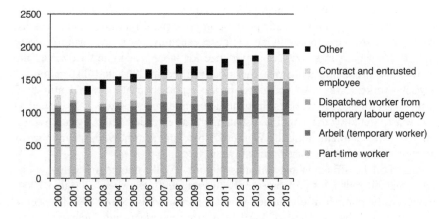

Figure 7.5 Development of various non-regular forms of work after 2000

Source: Ministry Internal Affairs and Communication, Statistics Bureau

Until 1985 any form of leasing or dispatching of workers was strictly speaking legally not permitted, although private work agencies already operated before 1985. Then, in 1986, the Japanese labour law was revised and new legislation, the Worker Dispatching Act, passed parliament, which for the first time officially allowed dispatching for a limited number of professions.[28] This approach was called a 'positive list' approach. In the following years, especially after the bursting of the bubble economy, growing pressure for further liberalization came from employer associations like the Japan Federation of Employers' Association, *Nikkeiren*, and other employer-friendly interest groups. In 1995 *Nikkeiren* in a study called 'Japanese Management in a New Era' demanded a shift in the approach from a positive list to a negative list, thus principally allowing agency workers for all professions with only some few exceptions.

The amendment of the Worker Dispatching Act of 1999 sought to combine both approaches. First the number of professions for which agency work was permitted was increased considerably. On the other hand, the negative list was also maintained, with professions that were strictly forbidden from agency work. Regarding the 26 professions on the positive list, restrictions on the duration of dispatching were lifted. For all other professions, which were neither on the positive or the negative list, agency work was in principal permitted on the basis of three conditions. These three conditions were: (1) agency workers could be assigned only temporarily, (2) the assignment of agency workers had to be in reaction to labour shortages that could not have been foreseen, and (3) agency workers must not replace ordinary employees. These conditions were set up with the intention that these new forms of employment '*maintain harmony with long established employment habits in Japan*' (Okamura Mihoko 2009: 125). In order to achieve this goal, the time permitted for the dispatching of agency workers was limited to one year.

Therefore, with the 1999 amendment to the Worker Dispatching Act, dispatching of agency workers was made possible in principle for almost all professions, with only a few exceptions. One of these exceptions was the manufacturing industry, because of its size and importance for Japan's economy. However, the ban on the producing industries was finally lifted in 2004 and dispatching of agency workers was also permitted in the manufacturing sector. Until 2007 dispatching of agency workers in manufacturing companies was also limited to one year. After 2007 the length of dispatching was extended to three years. Within 30 years after implementing the Worker Dispatching Act in 1986, the number of agency workers has increased almost twentyfold.

Based on the official labour statistics (Figure 7.5), the number of agency workers remained rather insignificant until the year 2000. Between 2000 and 2004, however, the number of dispatched workers increased to around 850,000 people and, with the abandonment of the ban on the manufacturing industries, their number had almost doubled to 1.4 million by 2008 – or roughly 3 per cent of the workforce. However, during the financial crisis, the problems of labour market liberalization became obvious. In winter 2009 it was first and foremost agency workers who were laid off. Within only a few weeks the number of laid-off agency workers skyrocketed, which all of a sudden brought the problem of precarious employment to public consciousness; many of these agency worker who had lost their jobs, and with that often also their homes, gathered in public parks and turned these into their dwelling areas, the so-called 'agency worker villages' (*haken mura*).[29] In reaction to the *haken mura* problem, legislative bills were brought forward to reintroduce the ban on agency workers in the manufacturing industries again, because the manufacturing industry was – as could have been predicted – the industry that absorbed the most agency workers and provided the least stable employment. However, there was no majority for an entire ban of agency workers in the producing sector. Instead, on October 2012 a law was passed in parliament forbidding the dispatching of day labourers, defined as persons working on contracts shorter than 30 days. With this at least the most precarious forms of agency work were eliminated.[30] Based on the official employment data, the number of agency workers dropped to 900,000 in 2012. However, most recently the number of agency workers is again growing and reached a level of about 1.2 million in the last two or three years.

However, the real dimension (and problem) of agency work is not displayed in the official employment figures. The official employment figures just indicate how many people at a certain time are employed as agency workers. However, agency work is much more volatile and much more complex. Data collected by the dispatching agencies merely indicate that just the pool of agency workers, which means the number of people available for dispatching, is up four times higher. But the actual number of agency workers is much higher, because most of them are not continuously working, but jump from job to job, and therefore at any point in time there is a considerable number not in work. Based on the data of the dispatching agencies, the numbers of agency workers are as shown in Figure 7.6.

From the data of the dispatching work agencies one can see that the financial

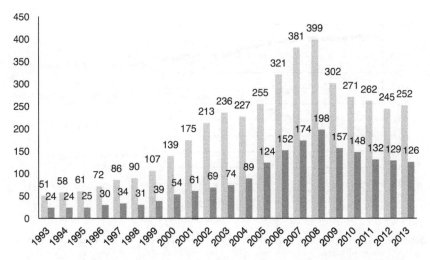

Number of available agency workers *Number of actuallly dispatched agency workers*

Figure 7.6 Development of number of workers available for agency work and number of actually dispatched agency workers (10,000 people)[31]

Source: Ministry of Health, Labour and Welfare: Accumulated Date Results of the Reports of the Work Agencies for Dispatching Workers

crisis in 2008 was at least a turning point in Japan, at least temporarily. As indicated above, it is obvious that, even before the regulation on day labourers was implemented, many companies seem to have realized that economic advantages achieved by employing an ever-increasing number of agency workers is at a certain point in time offset, and turned into an economic disadvantage, by a loss of experience, commitment and positive input attributable to an ever-shrinking number of regular employees. So the financial crisis, which as explained above had in other areas almost passed Japan by, might here have even led to a useful reconsideration of the merits and demerits of agency workers.

Turning back to overall development of the market for non-regular employment, one problematic aspect of the shift from regular to non-regular types of work especially is that the number of female employees who work, whether through choice or through lack of adequate regular employment, in non-regular forms of employment is continuously growing, as Figure 7.7 shows.

As in other countries, in Japan the majority of non-regular, especially part-time or temporary employees, are women. However, when looking at the development of female employment, we see that not only have non-regular forms of work considerably increased, but that the proportion of regular and of non-regular forms of employment have even completely reversed. Since around 2002 the majority of working women are no longer employed permanently but work part time, on time-limited contracts or as dispatched workers form work agencies.

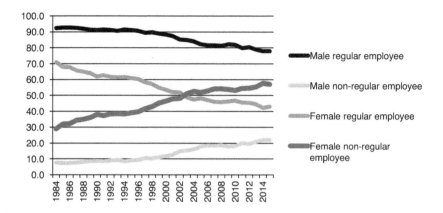

Figure 7.7 Development of regular and non-regular forms of work for male and female
employees

Source: Ministry Internal Affairs and Communication, Statistics Bureau

When looking at the proportion of male and female non-regular workers, we
see that the relation is about one male to two female employees and that this relation stays very constant (Figure 7.8).

With respect to female employment it is, first, especially noteworthy that the
tendency towards non-regular forms of employment is observable across all age
groups. Figure 7.9 displays this development with data for the years 1990, 2001
and the first quarter of 2015.

Second, it is noteworthy that, despite the fact that the overall proportion of
women in the workforce until the first quarter of 2015 was still constantly growing, there is an uneven development between regular and non-regular forms of
work. The absolute number of working women in Japan increased between 2001
and 2015 by 2.82 million from 20.76 million to 23.58 million. However, in the
same time period regular employment decreased by 850,000 jobs, while non-
regular employment grew by 3.74 million. Third, even in age groups where
regular employment is still increasing, non-regular employment is still faster
growing. For instance in the age group of women between 35 and 44 years of age,
regular employment increased by 30 per cent, while non-regular employment
grew by 43 per cent. In the two highest age groups, this is even more evident. In
the age group of women between 55 and 64 years regular employment grew by
10 per cent, and in the age group above 65 years even by 73 per cent. However,
non-regular employment increased by 61 per cent and by 317 per cent, respectively. And, fourth, even in the two youngest age groups where, due to the
demographic change, overall employment is decreasing, the decline in regular
employment is higher than the overall decline in the labour force of women in
these age groups. In the age group of women between 25 and 34 years of age, the

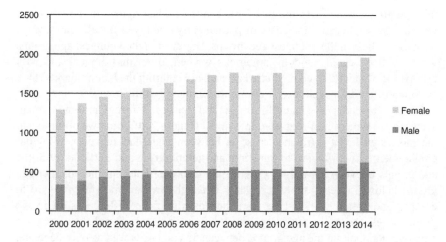

Figure 7.8 Proportion of male and female non-regular employees after 2000

Source: Ministry Internal Affairs and Communication, Statistics Bureau

overall decrease in the labour force between 2001 and 2015 was 18 per cent. However, regular employment shrank by more than 19 per cent, while non-regular employment only by 15.5 per cent. The same holds true for the youngest age group between 15 and 24 years of age, where the decline in regular employment was almost 40 per cent; this far exceeded the overall labour force decline in

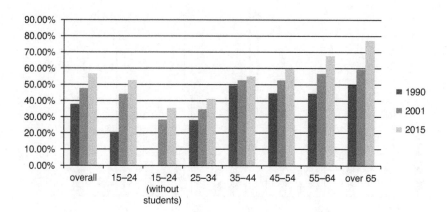

Figure 7.9 Development of non-regular female employment in percentage of respective age group

Source: Ministry Internal Affairs and Communication, Statistics Bureau

this age group, which was only 31.5 per cent.[32] The last figures in particular are very indicative, because the share of permanently employed females has always been the highest in the youngest age groups. But even in the youngest age groups where well educated, modern, ambitious women, often multilingual and with experience of having lived or studied abroad, are entering the labour market, they are often offered jobs that by far do not match their qualifications.

Finally, consider the data for male and female non-regular employment according to type of non-regular employment (Figure 7.10). First we observe that the most significant difference in scale between male and female non-regular employment is with regard to part-time and temporary work, particularly in the age groups between 35 and 64 years of age. Above 65 and in the younger age groups below 35 years, the discrepancy is not that remarkable. With regard to contract and agency workers, there are the least significant differences. In no age group does male non-regular employment exceed female non-regular employment. Only in the age groups between 55 and 64, as well as over 65 years, is the number of male contract and agency workers higher than the proportions of females in the same categories, as the graph below shows.

The data, however, give no indication as to whether the respective person works voluntarily on a non-regular employment contract or whether the person has no other choice because he/she cannot find a regular, permanent employment position or because there are no such offers on the job market.

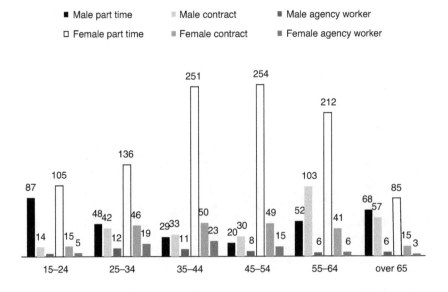

Figure 7.10 Male and female non-regular employment according to age groups in 2014

Source: Ministry of Health, Labour and Welfare

According to the most recent labour force survey of the Statistical Bureau of Japan, roughly 70 per cent of all female non-regular employees work in their current position because: (1) of the convenient work time (26.3 per cent); (2) the aim of working is earning additional money for the household budget or tuition fees for their children (25.5 per cent); or (3) there is a wish to combine household work, child education or care for elderly family member(s) with some commercial work (16.3 per cent). Some 13.6 per cent of non-regular female employees, however, give that they could not find an adequate regular job as the reason for working in the present position. With regard to male non-regular employees, the most often-stated reasons are: (1) that there is no regular employment available (27.9 per cent); (2) that they appreciate the convenient working time (22.7 per cent); and (3) that they have the opportunity to apply specific skills or knowledge in their present non-regular position (13.1 per cent).

When looking, however, specifically at the reason that there are no adequate regular jobs for those currently non-regularly employed, one observes considerable differences between males and females according to the form of non-regular work, as Figure 7.11 shows:

First, the data show that, with the exception of entrusted workers, the percentage of male non-regular employees who state that they cannot find a regular job (as their reason for working as a non-regular employee) is by far higher than the percentage for females. The lack of regular full-time positions is often especially stated by agency workers and contract workers as a reason for having to work on a non-regular employment assignment. Almost half of the male and more than a third of the female agency workers, as well as roughly one-third of both male and

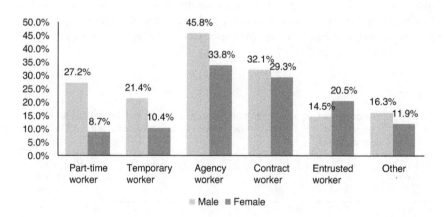

Figure 7.11 Lack of regular employment positions as reason for current non-regular employment in per cent for 2014

Source: Ministry Internal Affairs and Communication, Statistics Bureau: 2014 Labour Force Survey – Detailed Summary

female contract workers, give this reason. This is not surprising since both forms of non-regular work presumably provide the most precarious employment.

Despite the fact that a much larger percentage of male than female non-regular workers are dissatisfied with their current form of employment, because of the larger proportion of female non-regular employees the absolute number of women who are not able to find a full-time regular job is, however, with 1.81 million, slightly higher than the absolute number of men, with 1.76 million.

The limited supply of full-time jobs is also definitely one reason why people have to work longer in non-regular employment arrangements. In 2009, male employees worked on average 4.1 years in non-regular jobs, while female employees worked in this way one year longer, for 5.1 years. Compared with the figures for 1990, this was 1.1 years longer for men and 0.6 years longer for women. While only 21 per cent of men continued non-regular work for more than five years, 35.8 per cent of the women did so (Ministry for Health, Labour and Welfare, 2009: 9).

Before we finally look at income discrepancies, we have to define what part-time or non-regular employment means in terms of working hours. As Table 7.5 shows, part-time work is a flexible work arrangement that does not necessarily mean reduced working hours overall. Many employees work as long as, or even longer than, regular employees.

The table below shows the working hours of part-time employees for all industries. In some industries, however, like for instance the manufacturing industry, working hours are much longer. According to the same source, in the manufacturing industries average part-time working hours in 2009 were 30.7 hours; but 39.3 per cent of all part-time employees work longer than 35 hours, and 27.4 per cent work even longer than 40 hours per week. Hence the distinction is not so much working hours but rather the work contract, which of course provides less favourable clauses than the contract for ordinary employees. So, besides labour flexibility, reduction of labour costs especially provides the highest incentive for the employing company to employ non-regular employees instead of regular ones. This is shown in Table 7.6.

Table 7.5 Working hours of part-time employees for all industries in 2009 as a percentage share of all part-time employees as well as average weekly working hours

Weekly working hours	1–14	15–29	30–34	35–39	40–48	Over 49	Average working hours
All part-time employees	15.6%	43.7%	13.3%	8.5%	15.3%	3.5%	26.5
Male	15.3%	32.2%	12.8%	7.4%	24.4%	7.4%	29.8
Female	15.5%	46.9%	13.5%	8.8%	12.7%	2.5%	25.5

Source: Ministry of Labour, Health and Welfare: *The Current Situation of Part-time Work*, p. 8

Table 7.6 Income discrepancies between regular and non-regular employees in
10,000 yen

| Age | Regular employee | | | Non-regular employee | | | Income |
	Monthly salary	Yearly bonuses	Yearly income	Monthly salary	Yearly bonuses	Yearly income	difference
up to 17	15.4	3.9	188.1	12.5	0.4	150.4	37.7
18–19	18.6	12.5	235.7	16.9	2.0	204.8	30.9
20–24	22.3	38.1	305.2	18.9	8.2	235.0	70.2
25–29	26.9	68.9	391.7	21	11.5	236.5	128.2
30–34	31.5	87.8	466.3	22.1	13.1	278.3	188.0
35–39	36.8	113.0	554.0	21.4	15.1	271.9	282.2
40–44	40.4	133.5	618.7	20.4	15.2	260.0	358.7
45–49	41.5	138.7	637.1	20.5	18.2	264.2	372.9
50–54	41.8	137.7	639.5	19.8	18.2	255.8	383.7
55–59	29.9	123.2	602.4	20.6	20.3	267.5	334.9
60–64	31.5	70.2	447.8	23.2	37.9	316.3	131.5
65 plus	28.7	51.3	395.4	23.2	18.8	252.8	142.6
Average	35.0	103.7	523.5	19.5	17.4	267.0	256.5

Source: Ministry of Internal Affairs and Communication, Statistics Bureau

As Table 7.6 indicates, the wage differences between regular and non-regular employment are especially significant in the age groups of employees between 30 and 60 years of age.[33] This large discrepancy in income, however, stems not so much from the difference in monthly salaries, although this difference is also not marginal, but especially from the difference in yearly bonus payments. Although, particularly after the bursting of the bubble economy, bonus payments were considerably reduced, still the discrepancy that results from bonus payments is considerable.

Summarizing the outcome of the liberalization of the Japanese labour market, particularly after 2000, it is appropriate to state that the labour market is not just divided into regular, long-term employment and non-regular employment. This division already existed in the past. In fact, as a result of the reforms, the labour market has been split up into various categories – spanning a spectrum from permanent, lifelong employment on the one side, to part-time, temporary, contract and post-retirement work, as well as other forms of occasional work, and precarious forms of work like dispatched work or even agency day-labourer work, on the other.

Although the first labour market reforms had been implemented before the bursting of the bubble economy, the reforms after 1999 went hand in hand with the reforms of company law and the corporate governance system. To put it simply, to change the company system it was necessary to liberalize the labour market, and to change the labour market the company system had to be altered. The two reforms cannot be separated from each other; they simply depended on each other. The traditional Japanese company system almost completely relied on an internal labour market. The only external labour market was for high school and university graduates. Once they were taken into the companies, the only way on the human resource mamnagement side to improve effectiveness, efficiency, productivity and profitability was to educate the employees and enhance their knowledge and abilities. There were no real external labour markets. Non-regular employment was never a challenge to regular employment. The labour market reforms, however, created external labour markets that are increasingly challenging the internal labour market, especially for simple tasks in the manufacturing and service industries. The problem with this is that the permeability between the external and internal labour markets is one-sided. It is very difficult to get back into the internal labour market, once one has been out. Especially for higher aged non-regular workers, it is almost impossible to get back onto the track of regular, permanent employment.

4   Summary and outlook

Japan in the last two to three decades has departed considerably from its 'traditional' post-war management and company system. The institutional conditions for these changes were provided, on the one hand, by the new company law that was after a long period of time finally established in 2006 and the codification of a corporate governance framework that became the standard for listing and evaluating companies on the Tokyo stock exchange. These reforms were implemented in order to react to the bursting of the bubble economy, which was solely blamed on the traditional company system, to enhance transparency and to comply with international standards of corporate governance.

On the other hand, and reacting to pressure coming from employers' organizations, from 1986 labour market regulations were successively implemented in Japan with the intention of liberalizing the labour market and increasing labour flexibility. In fact these labour regulations legalized agency work, which already existed before, albeit in a legal grey zone, and established a low-wage sector, initially for a limited number of professions, but later in almost all professional fields, including especially manufacturing.

Both reforms, as we have seen, altered the post-war Japanese company and management system considerably. They in particular brought an end to cross-shareholding between companies and banks belonging to a group of companies, and they created an outside labour market, which did not previously exist in this form in Japan.

Assessing the result of these reforms more critically, one has to state first that the new company law and the corporate governance regulations did not prevent several company scandals from happening, like (for instance) the Olympus scandal in 2011/12 over concealment of huge investment losses and allegedly close connection to 'anti-social groups' (as the Japanese mafia or *Yakuza* is officially called). Second, one has to state that the labour market reforms led to precarious work forms, to a working poor and, given the still existing rigidities of the Japanese employment system, to a growing number of people who, once out of regular employment, never find their way back to regular forms of employment.

For these tendencies hard data can also be given as evidence. Since the 1970s the income-based GINI index for Japan is continuously rising and is now well above the average of the OECD countries.[34] Already 47 per cent of all social aid recipients are elderly people with no income or too low incomes from pensions. It can be expected that their number will increase dramatically in the future, when more people who never or only temporarily worked in regular forms of employment will retire from working (data: Ministry of Health, Labour and Welfare, Survey on Recipients of Social Aid, Nov. 2014)

However, long established systems do not change completely overnight. Most especially, the world financial crisis of 2008 showed that 'adjustment to international standards' is not necessarily providing a solution to the task of how to cope with a changing economic environment and with growing competition outside Japan. Interestingly, however, in the midst of this crisis, the Japanese traditional system still proved to be strong and relevant. Despite the heavy economic downturn, the companies maintained the employment level of their core employees. There was almost no increase in unemployment registered amongst regular workers. With additional training programmes or short-time work, only partially paid or non-paid vacations, etc., companies managed to keep their employees during the height of the crisis.[35] Cooperation between the 'social partners' still worked. For instance in the car industry the social partners agreed upon the introduction of a new taxation system and an incentive scheme for purchasing ecologically friendly cars.

Also new proposals for regulation of the agency worker labour market that intend to restrict agency work for the same position to three years, by this means enhancing the return to regular employment, indicate that the social actors are trying to find a balance between reacting to new global challenges and protecting social coherence. It is, however, questionable, whether a limitation to three years will indeed increase permanent employment again.[36]

And finally, some of the most successful companies in Japan still maintain distinctive features of the traditional Japanese management and company system. Toyota, for example, maintains independence from external financing and keeps financially close ties with other companies belonging to the Toyota group. With this, the company is not only the most profitable car manufacturer in the world, but also the most highly valued when measured on the basis of share prices.

Japan, like all advanced economies, will also in the future face challenges from emerging countries, especially from nearby Asian countries that all have very specific socio-economic models, not necessarily matching the 'international' (in other words, the American) standard. In addition to that, Japan will also have to find innovative solutions in order to meet the challenges of a rapidly aging and also shrinking population. This will demand reforms of the labour market, not just opening up a low-wage labour market sector. And it further demands that companies come up with new ideas to provide opportunities in particular for females to combine family and work, and to provide female employees with equal opportunities to progress to the highest management positions. It will be particularly necessary to react to the demands of the young generation, the so-called 'millennials', who have different attitudes towards work and for whom loyalty to a chosen company is not a highest priority. And, finally, Japanese companies will have to come up with new approaches to the issue of work–life balance, not only for female and younger employees, but for all employees.

These labour market- and employment-related issues will, of course, and to a large extent, depend on the question as to whether Japan's economic as well as management and company system will continue to proceed on the way from 'stakeholder' to 'shareholder' capitalism. This process means that income distribution continues to shift towards the capital side; it remains to be seen whether the pendulum will to some extent swing back in favour of broader social inclusion and an income distribution that is not disadvantageous towards the labour side. This will also relate to the question as to how Japanese companies in the future define their corporate social responsibilities. But there will be no way back to the old Japanese management and company model, often described as 'Japan Inc.'

Japan and the Japanese companies will have to find new ways in order to meet the challenges ahead. First and foremost amongst these is a changing demographic structure, and the changing values and attitudes of an ever smaller younger workforce at home, all with increasing international competition abroad – especially from dynamically developing regions in East and Southeast Asia.

Notes

1   What was named 'lean production' in the West was a production system that evolved in Japan after the Second World War. At that time, sales numbers were too low to allow Japanese car manufacturers to implement mass production systems like in the USA or Europe. Instead, their approach was to decrease costs by eliminating all kind of waste in production and to increase production flexibility. Over time a production system evolved that allowed Japanese manufacturers to produce a large variety of car models in small quantities. Still most representative of this way of production is the Toyota production system, which serves as a benchmark, not only for the automobile industry, for effective and efficient production.
2   Because of the sluggish economic development after the bursting of the bubble economy, the years after 1995 are often referred to as the 'lost 10', recently even as the 'lost 20', years.
3   The bubble economy resulted from an overheating of the real estate and stock markets. As a result of the bursting of the bubble, the NIKKEI stock market index almost collapsed, from 38916 yen in December 1989 to 20984 yen in September 1990 and to 15910 in July 1992 (Data: Tokyo Stock Exchange). With the stock market, the real estate market also broke down, leaving the banks with estimated bad loans amounting to about 30 per cent of Japan's GDP (Data: OECD; Worldbank).
4   Consumer prices between 1999 and 2005 fell by 0.45 per cent and between 2009 and 2012 by 0.6 per cent.
5   The figures displayed in the graph are percentage growth in GDP and are based on the Japanese fiscal year. The Japanese fiscal year starts on April 1st and ends on the March 31st the following year. The thick horizontal line indicates the average growth rate for the respective period.
6   Japan's population is quickly aging. Low birthrates and the highest life expectancy in the world have resulted in the fact that the population in the working-age band between 15 and 64 years is constantly shrinking. Since 2009 the population as a whole has also been declining.

7   Since the data above are based on Japan's fiscal year (FY), the economic slump was the fiercest in 2008, which includes the first quarter of 2009, which showed the deepest downturn.

8   Much more than the financial crisis, the North-East Japan earthquake of March 11th 2011 affected the level of employment negatively. The decrease in 2011 is, however, also attributable to the flood disaster in Thailand that disrupted supply networks between Japan and Thailand.

9   According to data of the Ministry for Internal Affairs and Communication, small enterprises (up to 29 employees) employed 15.9 million people, medium-sized enterprises (between 30 and 499 employees) had 18.7 million people on the payroll, and 14.3 million are working at large enterprises (over 500 employees) www.stat.go.jp/english/data/roudou/.

10 Of course there is quite a large percentage of hidden unemployment in Japan. So, including hidden unemployment, the real unemployment rate might be around 5 to 6 per cent.

11 ASEAN, the Association of Southeast Asian Nations, was founded in August 1967. Today ASEAN has ten member countries, Brunei, Cambodia, Indonesia, Laos, Malaysia, Myanmar, the Philippines, Singapore, Thailand and Vietnam.

12 As an example, the electronic industry can be given, where Japanese companies are continuously losing market share or even entire market segments to their Korean competitors; or, regarding the auto industry, where Hyundai is becoming Toyota's fiercest competitor in the North American market.

13 For a more detailed outline of Japan's traditional company and management system, see Dore (2010).

14 There had been always forms of peripheral employment like part-time work, time-contract work, female work, etc. in Japan.

15 Differently to Germany's 'social market economy', this post-war Japanese company system was never codified in Japan's company or labour legislation. In fact Japan's company law is not much different from the legislation in the USA.

16 The two other arrows were (1) monetary expansion and (2) fiscal and investment stimuli.

17 According to estimations of the Organisation for Economic Co-operation and development (OECD) and the World Bank, just resolving the bad loan problem involved costs amounting to 30 per cent of Japan's GDP, and additional expenses, which arose from public costs related to dealing with the problems, accounted for another 9 per cent of GDP.

18 Specifically the FSA received the right to (1) demand reports and materials concerning the business or financial conditions of a bank (including its agencies and subsidiaries), (2) conduct on-site inspections at bank premises, (3) penalize misconduct that could lead in extreme cases to suspension of a bank's operations or revocation of its licence and (4) order a bank to hold a part of its assets within Japan.

19 There are two categories of regional banks, one for prefectural and one for smaller banks that belong to the Second Association of Regional Banks.

20 Data indicate percentage distributions of market value owned by type of shareholder. Insiders are basically city and regional banks, life and non-life insurance institutions and business corporations. Outsiders are investment and pension funds, individuals and foreign shareholders. The share of foreigners is also included in the data displayed for outsiders. In order to indicate the dimension of the rising share of foreign shareholders, however, the data for foreign shareholders are also separately displayed.

21 In fact there was much talk about Japan's classless society, also because in surveys a vast majority of people perceived themselves to belong to the middle class.

22 Salaries (in yen) until 1985 are for male graduates working in administrative functions. From 1990, salaries are for both sexes. According to the Statistical Bureau of Japan, in 2011 the enrolment rate in high schools was 98.2 per cent and in colleges

and universities 53.9 per cent of the respective age groups. The data for new university graduates are relevant because they are the core employees, on which the Japanese company system has for long been based, the markets for university and high school graduates being the only external labour market in which the companies fiercely competed to get the best talents.

23 The bonus payments are especially relevant with respect to the real estate market and the market for durable goods like automobiles or furniture, etc., since bonuses are often used to purchase cars or to pay back loans for real estate, cars or other expensive products.

24 Presumably, however, over the same time period of two decades, remuneration of top managers in the USA and Europe has increased even faster than in Japan. For instance it is very well known that all members of the board of directors of Toyota together earn less than just the CEO of Volkswagen, when performance-related compensation is included.

25 Other forms of non-regular work like, for instance, temporary or occasional work done by high school or university students of course have always existed in Japan. However, student work was never an issue and is also disregarded in this chapter.

26 Entrusted employee usually refers to senior employees, who are rehired by the former employer after retirement or transferred from (usually) a higher ranked company to a lower ranked one within a company group. Contract workers are usually specialists assigned for performing a task over a certain time period.

27 These definitions are by no means exactly fixed and in fact vary even within different government institutions which are responsible for publishing labour market surveys.

28 Almost at the same time the Equal Employment Opportunity Act passed parliament, which should strengthen the rights of female employees. The coincidence of these two legal acts, however, can also be interpreted in that the legislator was quite aware of the fact that the Worker Dispatching Act will especially affect female employment. The implementation of the Equal Employment Opportunity Act might therefore be perceived as an act to provide a counterbalance.

29 The number of unemployed agency workers is generally considerably higher than the number of unemployed regular or part-time workers. However, in 2009 the percentage of unemployed agency workers jumped from 7.9 per cent in 2008 to 22.2 per cent, while the unemployment figure for part-time workers only increased from 2.9 per cent to 3.6 per cent and that for regular workers from 1.7 per cent to 2.4 per cent (Statistics Bureau: 2010 Labour Force Survey – Detailed Summary).

30 This legislation led to a considerable reduction in agency workers in the producing industries. A human resources manager of a company affiliated to Toyota said to the author at that time that the company would have abandoned agency workers anyway, because they had not contributed to improving the production system in the way that ordinary employees do. Obviously, however, it took the new legislation to support this insight before it finally found expression in adequate corporate decision making.

31 The figure for actually dispatched workers consists of people who have worked in a steady or non-steady position during the last 12 months, not considering how long the assignment has been.

32 The number for the age group of women between 15 and 24 years of age excludes school and university students, who usually pursue only temporary work.

33 The comparison is based on regular and non-regular full-time employment. Therefore it does not consider part-time or temporary workers, but in the first place contract, entrusted and agency workers who basically are working full time.

34 According to the Ministry of Health, Labour and Welfare, Japan's income-related GINI index increased from 0.354 points before redistribution in 1972 to 0.532 in 2008. Of course this increase is not solely attributable to labour market reforms, but has various reasons like the increase of single households or of elderly people, etc.

35 Also in this respect it is worth mentioning that in Germany the system of co-determination – which in the 1990s came under much pressure, not least from EU regulators – also provided the institutional basis for dealing with and overcoming the crisis, even leading to a renaissance of co-determination.

36 Currently (July 2015) new regulations for agency workers are being discussed in parliament that intend to generally limit agency work in a single position in one company to three years. The new proposals hold that an agency worker can only work longer than three years in the same position when he/she gets permanently employed by the work agency or when a committee of management and unions at the company where the agency worker is dispatched agrees on continuing the work contract with the work agency for this worker. It can be expected that the unions would push for changing the employment status to employment at the company, while management might just intend to substitute one agency worker with another one. It is most unlikely that the agency worker actually gets a permanent job at the work agency. So it is quite questionable whether the discussed new regulations will indeed have the effect that more agency workers get back into regular employment.

Bibliography

Buchanan, J. and Deakin, S. (2009) 'In the shadow of corporate governance reform: Change and continuity in managerial practice at listed companies in Japan', in Whittaker, H. D. and Deakin, S. (eds) *Corporate governance and managerial reform in Japan*, Oxford: Oxford University Press, pp. 28–69.

Dore, R. (2006) *Dare no tame no kaisha ni suruka?* (For whom do we have our companies?), Tokyo: Iwanami Shoten.

Dore, R. (2010) 'Afterword', in Hyun-Chin, Lim, Wolf, Schäfer and Suk-Man, Hwang (eds) *New Asias: Global futures of world regions*, Seoul: Seoul National University Press, pp. 281–301.

Goka, Kazumichi (2014) *Hiseiki daikoku Nihon no koyō to rōdō*, Tokyo: Shinnihon Shuppansha.

Matsuzuka, Yukari (2002) *Changes in the permanent employment system in Japan: Between 1982 and 1997*, New York and London: Routledge.

Miura, Mari (2012) *Welfare through work. Conservative ideas, partisan dynamics, and social protection in Japan*, Ithaca, NY and London: Cornell University Press. (See especially Chapter 4: 'Reforming the labor market'; and Chapter 5: 'Who wants what reform?').

Morinaga, Takurō and Kōga, Chūichi (eds) (2008) *Bukka no bunkashi jiten* (Encyclopaedia of the cultural history of prices), Tokyo, Tenbōsha.

Nakamura, Masao (2011) 'An overview of corporate governance reforms in post-bubble Japan: Institutional change and selective adaptation', in Holroyd, C. and Coates, K. (eds): *Japan in the age of globalization*, Abingdon and New York: Routledge, pp. 96–117.

Nikkeiren (1995) *Shinjidai no nihonteki keiei* (Japanese management in a new era), Tokyo, Nikkeiren.

Tarohmaru, H. (2011) 'Income inequality between standard and nonstandard employment in Japan, Korea and Taiwan', in Sato, Yoshimichi and Imai, Jun (eds) *Japan's new inequality. Intersection of employment reforms and welfare arrangements*, Melbourne: Trans Pacific Press, pp. 54–70.

Takurō, Morinaga and Chūichi, Kōga (eds) (2008) *Bukka no bunkashi jiten* (Encyclopedia of the cultural history of prices), Tokyo: Tenbōsha.

Sato, Yoshimichi and Imai, Jun (2011) 'Regular and non-regular employment as an additional duality in Japanese labour market: Institutional perspectives on career mobility', in Sato, Yoshimichi and Imai, Jun (eds) *Japan's new inequality. Intersection of employment reforms and welfare arrangements*, Melbourne: Trans Pacific Press, pp. 1–32.

Online*

Aoyagi, C. and Ganelli, G. (2014) *Unstash the cash! Corporate governance reform in Japan*, www.imf.org/external/pubs/ft/wp/2014/wp14140.pdf.

Freshfields Bruckhaus Deringer (2015) Japan: Landmark corporate governance reforms, www.freshfields.com/uploadedFiles/SiteWide/Knowledge/Japan%20-%20landmark%20corporate%20governance%20reforms%20(May%202015).pdf.

Goto Gen (2013) *The Outline for the Companies Act Reform in Japan and Its Implications*, http://ssrn.com/abstract=2348554.

Japan Institute for Labour Policy and Training (2011) *Non-regular employment – issues and challenges common to the major developed countries*. JILPT Report No. 10, 2011, www.jil.go.jp/english/reports/documents/jilpt-reports/no.10.pdf.

Keidanren (2015) *Shinkigaku sotsusha kettei shoninkyū chōsa kekka 2015.3* (Survey results of March 2015 regarding first salaries of new graduates), www.keidanren.or.jp/policy/2015/090.pdf.

Ministry of Health, Labour and Welfare (2009) *Pātotaimu rōdō no genjō* (The current situation of part-time work), www.mhlw.go.jp/stf/shingi/2r98520000011q6m-att/2r98520000011wjl.pdf.

Miyajima, H. (2014) *Perspectives for Corporate Governance Reform in Japan*, Research Institute of Economy, Trade, and Industry (RIETI), www.rieti.go.jp/en/special/policy-update/057.html.

Okamura Mihoko (2009) *Rōdōsha hakenhō kaisei mondai* (The problems of amending the agency workers' legislation), National Diet Library www.ndl.go.jp/jp/diet/publication/refer/200910_705/070506.pdf.

Tokyo Stock Exchange (2014) *Shareownership Survey 2014*. www.jpx.co.jp/english/markets/statistics-equities/examination/b5b4pj000000m5dr-att/e-bunpu2014.pdf.

* All internet addresses were last accessed on August 21 2015.

Databases

Bank of Japan: Statistics, www.boj.or.jp/en/statistics/pub/pim/index.htm/

Cabinet Office, Government of Japan: Statistics, www.esri.cao.go.jp/index-e.html

Ministry of Health, Labour and Welfare: Accumulated Date Results of the Reports of the Work Agencies for Dispatching Workers (since 1993); in Japanese, www.mhlw.go.jp/stf/seisakunitsuite/bunya/0000079194.html

Ministry of Health, Labour and Welfare: Basic Statistical Survey on the Wage Structure (several years); in Japanese, www.mhlw.go.jp/toukei/list/chinginkouzou.html

Ministry of Health, Labour and Welfare: General Survey on Dispatched Workers, www.mhlw.go.jp/toukei/list/40-20.html

Ministry of Health, Labour and Welfare: Survey on Living by Social Security; in Japanese, www.mhlw.go.jp/toukei/list/70-15.html

Ministry of Health, Labour and Welfare: Survey on Recipients of Social Aid (Nov. 2014) (in Japanese).

Ministry of Internal Affairs and Communication, Statistics Bureau: Labour Force Survey (several years), www.stat.go.jp/english/data/roudou/

8 China and the global recession

Jason Begley and Tom Donnelly

At the time of writing, China is poised to overtake the United States as the world's largest economy, five years earlier than previously anticipated (Hopkins, 2014). Such a position in itself is indicative of just how far China's economy has progressed since the introduction of the Open Door policy in 1978 when, after the death of Mao Zedong and under the leadership of Deng Xiaopeng, China cautiously opened her economy to the wider world. The motivation behind this was acknowledgement by China that its economy, technology, consumption levels and business organisation lagged severely behind not only that of the West but also in comparison with other emerging economies in East Asia. There was no choice but to seek external assistance in bringing about a programme of modernisation. In other words this was a policy of economic catch-up (Donnelly *et al.*, 2010). Just over two decades later China joined the World Trade Organization (WTO), a signal that it had almost integrated fully into the fabric of the world trading community (Hopkins, 2014; *The Times*, 2014). This is easily illustrated. Between 1978 and 2010 China's gross domestic product (GDP) expanded from contributing only 1.7 per cent in the world economy to 9.5 per cent when valued at market prices (Li *et al.*, 2010).

Over the period from 1990 to the early 2000s China enjoyed enviably high rates of economic growth, averaging circa 10 per cent per annum. Much of this was based on the pillars of domestic investment, inwards foreign direct investment (FDI), low cost labour, property development and an undervalued currency. The latter was of great importance to the high export growth of low-level, labour-intensively produced goods such as textiles, clothing and similar apparel. In recent years electrical goods such as data-processing equipment and medical instruments as well as machinery and capital goods have become more prominent in the export statistics. Equally it needs to be remembered that China imports large quantities of parts/components for incorporation into its own products and so intra-industry trade looms high in Chinese trade figures (Li *et al.*, 2012).

The result was that the Chinese were able to accumulate large balance of payments surpluses, denominated primarily in American dollars, which in turn were used by the Chinese authorities to purchase US treasury bills, bonds and equities so that by 2008 China's US dollar financial reserves stood at US $1.95 trillion. In effect, by purchasing US financial instruments, China was facilitating high levels of consumption in the American market, particularly in the housing boom of the 2000s (Nan, 2009). In such circumstances, of course, it requires one country to be prepared to tolerate a high deficit in its balance of payments and another to be prepared to participate heavily in financing it and keep the cycle turning.

The main purpose of this specific chapter is to discuss how China reacted to the global economic crisis, especially after the collapse of Lehman Brothers in 2008. There were some at the time who advocated almost a decoupling of the economies of East Asia from the global mess and focussing on increased intra-regional trade by way of compensation (Lardy, 2012). This would have been impossible for China, which by that time was deeply integrated into the world economy, especially after gaining WTO membership in 2001. It could not stand immune from the effects of the slowdown in economic activity in the United States and the European Union (EU) as these were by far the main destinations for Chinese exports (Lardy, 2012).

Before embarking on the remainder of this discussion, several points need to be made concerning China's economic, political and social structures. First, China is unique and should not be considered as a democracy in any precise Western sense of that word. Second, it remains a one-party state under the Communist Party whose view is to create a 'harmonious society' in which there is little separation between politics, economics and social welfare. Third, the Open Door policy of 1978 led to what is best considered as a form of economic dualism. Basically, the central government, while permitting and encouraging economic reform by introducing a degree of economic liberalisation into the state-owned enterprises (SOE) and also in the rural economy through the provincial town and village enterprises (TVE), as part of a wider attempt to mitigate regional economic imbalances, still retained a firm grip on overall policy direction. Sometimes this policy is referred to as the 'two hands' system or the 'visible' and 'invisible hands' system, which means that the state still determines overall direction of the economy while allowing a degree of decentralisation (Angang, 2010; Cai *et al.*, 2002).

Fourth, regional economic imbalances are only one of China's large economic problems. Such imbalances are compounded by too high a share of GDP being taken up by investment, particularly in industries such as steel and heavy industry, and too little in consumption or in the service sector generally. In the years 2006–2010 investment in industry and manufacturing reached over 50 per cent of GDP with private consumption reaching only 34 per cent between 2008 and 2010, by far the lowest share in any major economy. Indeed, in the United States the comparable figure is circa 70 per cent, lying perhaps too far in the opposite direction (Lardy, 2012). Bound up with low levels of consumption are comparatively low levels of expenditure on health, medical care and education. There are pressures to improve living standards through increased expenditure and the social welfare of people on low incomes in rural communities through improving the level of transfer payments (Xiaonian, 2010). As Guthrie (2009) notes, this situation is not helped by an ageing population and the consequences of the one-child policy on demographic structures. Fifth, pressures emanate from increasing urbanisation, which is exacerbated by internal migration by younger workers, often in their twenties in search of employment. Estimates of the scale of the migrant workforce vary, but Yang and Huizenga put the figure at 170 million. Migrant labourers do not enjoy full rights when moving to the cities, as often access to urban social security, health care, housing and education is limited as a result of their inability to obtain official city residential permits, and so living conditions are frequently poor (Yang and Huizenga, 2010). Lastly, recent urban growth in China has been extremely fast. There are now over 150 cities with more than 1 million residents and, due to weaknesses in the country's energy and waste disposal policies, environmental pollution is high, demanding high levels of governmental intervention at both national and provincial levels (Zhenzhong, 2010).

As part of its drive to work with foreign investing enterprises (FIE), the Chinese government realised that it had to implement legal reforms that would ease its shift from a totally planned economy to market capitalism. Essentially, these were designed to permit the state to slowly loosen its controls on the economy and so avoid the 'shock therapy' route that the United States had proposed when advising East European countries when they were making similar structural transitions. In other words, there was no push for mass privatisation. Equally there was an encouragement of inwards FDI, but only under conditions laid down by the Beijing government on relative shareholdings in joint ventures, partnerships and location (Donnelly *et al.*, 2010). In line with opening its borders, the Chinese legal system was altered as part of wider institutional reform to support the newly emerging economic order. None of this was rushed; it was carried through systematically over the 1980s and 1990s (Guthrie 2009; Donnelly *et al.*, 2010). This was a slow, incremental approach to change which allowed the government to act as a stabilising element in both economy and society. For example, firms gained experience in one area such as setting prices, competing for contracts and producing more efficiently, and then further progress in operational liberalisation could effectively be allowed. The state's planning role gradually reduced, thereby pushing business and managerial responsibilities further down the political/economic chain to managers and provincial officials, but with this came an accompanying accountability for decision making, actions and outcomes (Naughton 1995; Rawski, 1994).

Against the above contextualisation, the remainder of this chapter will focus first, and very briefly, on the basis of China's swift rate of economic growth. Second, it will then examine how the financial crisis impacted on China's exporting industries before, third, trying to evaluate how effectively the Chinese authorities responded to the ensuing economic difficulties through its monetary and fiscal policies. Lastly, the chapter will attempt to draw conclusions and see whether any lessons can be learned from the Chinese experience.

The impact of the financial crisis of 2008

Beginning in the United States in 2008, the international recession spread to the European Union and then to other parts of the world. Space precludes a detailed discussion of the origins of this event and, likewise, this means that in dealing with its impact on China a degree of selectivity is essential. The recession's impact varied in timing and process from country to country, and so individual country responses differed to the degree that countries emerged from the nadir of recession at different times (Lardy, 2012).

The consequence of the economic crisis was a reduction in global economic activity, with credit being tightened. Between 2008 and 2009 the pace of world GDP growth fell from 5.4 per cent in 2007 to 2.8 per cent in 2008, with output shrinking by another 0.6 per cent in 2009. Naturally, this was reflected in a contraction in international trade, which had been growing at circa 9 per cent per annum; it slowed to 2.5 per cent in 2008, before falling by circa 10 per cent the following year (Fardoust *et al.*, 2012). These figures were further mirrored in falling investment levels as a result of the existence of excess capacity, which raised the spectre of weak profitability and excessive risk. The outcome was rising unemployment through factory closures, and the ensuing redundancies in the United States and the EU contributed to growing public sector indebtedness in many countries There was no respite until 2010 when a fragile recovery set in (Elwell, 2013; Williams, 2012).

China's financial sector came off reasonably lightly in comparison with America and several EU states. Whereas the impact in the last mentioned geographical areas hit their banking and financial sectors, in China it was primarily the 'real economy' that suffered (Zhang *et al.*, 2010). This was because the Chinese financial system had little exposure to the risks on the world financial markets, particularly those linked to highly complex financial instruments. Moreover, the Chinese government exercises firm controls on the individual's right to invest savings outside China. The majority of Chinese outward investment flows are under the aegis of state banks and SOEs, whose policies tend to be conservative rather than risk-taking when operating in international markets. Indeed, the biggest financial loss in the crisis was that of US \$2billion sustained by the Bank of China following the American subprime crisis through investments in Freddie Mac and Fanny Mae (Yang and Huizenga, 2010). Nonetheless China was impacted by the recession. China's economic growth in 2008 was a mere 8–9 per cent, low by China's recent historical standards, in comparison with 13 per cent the previous year. The growth rate of industrial production dropped to 8.2 per cent, representing only circa 50 per cent of the rate in the previous year. China's stock market boom of 2005–2007 came to an end. Share prices plummeted by 70 per cent, with real estate values and house prices stagnating before falling, and even then consumers preferred to invest in property rather than in shares or the banks because of potential higher rates of return in the medium to long term (Li *et al.*, 2012; Yongding, 2009). In the case of some manufactured products the rate went from double-digit growth levels to negative. In the first quarter of 2009 growth fell as low as 6.1 per cent before any sign of an upturn appeared (Zhang, 2009). Lastly, China's net FDI inflows fell to US \$122 billion in 2008 and to US \$70 billion in 2009, dropping 15 per cent and 42 per cent year on year respectively, before recovering in 2010 (Li *et al.*, 2012; Morrison, 2009).

The roots of some of China's external problems lay in the Beijing government's decision to pursue an export-oriented strategy in 1988. A major objective of this policy was not simply to obtain higher rates of GDP growth through foreign trade, but to create employment and so absorb the increasing number of rural–urban migrants in labour-intensive activities. A number of mainly coastal regions were granted greater degrees of economic autonomy and were encouraged to develop overseas trading links. Included in this group were Shandong, Beijing, Hainan, Guangxi and Shandong, for example. The precedent for this policy was the success achieved in other countries such as Taiwan, South Korea and Japan, all of which had prospered through the development of export-driven economies (Guthrie, 2009). Additionally, it was hoped that, with experience, economic competitiveness would improve, firms could eventually move up the value chain and so much-needed foreign currency would be raised to build up financial reserves. To a degree the strategy has proved successful in that it first provided circa 80 million jobs, with 28 million of these employed by FIEs (Morrison, 2009), and so enmeshed China into the mainstream of global trade. It has helped to entice foreign firms into these areas and so generated further employment, in that firms from South Korea and Taiwan, as well as US firms such as Dell, Motorola and Walmart, have set up production units to take advantage of relatively low-cost labour, also contributing to the growth of China's large trade surpluses. In other words, a significant portion of China's exports can be accounted for by the activities of multinational firms. However, it left parts of these areas overdependent on a narrow range of what might be termed 'staple industries', with little alternative sources of demand or employment and, therefore, vulnerable to any major downswing in the international economy. (Li *et al.*, 2012; Guthrie, 2009).

China's problem was that the contraction in world trade impacted seriously on the external account, a reflection of the overdependencies outlined above. Export growth was 34 per cent in 2004, 23.2 per cent in 2005, 23.6 per cent in 2006 and 23.5 per cent in 2007. In the following year, 2008, growth fell significantly to 17.2 per cent, the lowest for five years, with the downward trend continuing into 2009 (Yongdin, 2009). Part of the problem lay in so many of China's exports being directed at G3 countries, the United States, EU and Japan, which until 2008 accounted for approximately 46 per cent of exports. With the G3 falling into recession and the financial crisis, it is little wonder that export-dependent firms suffered badly. Similarly, other outlets that were considered as alternative sources of demand, such as Hong Kong, Taiwan, South Korea and the members of the Association of Southeast Asian Nations (ASEAN), also felt the cold wind of change and so their rates of GDP growth faltered, further weakening sources of external demand. Lastly, because of the fall in world energy and commodity prices in the recession, and also as a result of the consequent declining internal demand in Chinese industry generally, imports decreased quicker than exports, which, ironically, in turn caused the trade surplus to increase. Finally, with such a high proportion of its overseas reserves being held in US dollars, any threat of dollar devaluation would have seriously eroded their value, with major implications for China's financial position in the world economy (Zhang, 2009).

A direct consequence of the deteriorating trade situation was the bankruptcy of a large number of enterprises in the previously buoyant coastal regions. In Guangdong over 600 firms went out of business. The Hong Kong economy, too, was relatively exposed to the crisis because of its specific exposure to being an international commercial trading entrepôt. Between September 2008 and March 2009 the volume of trade dropped by 21 per cent, with output in the financial sector, too, being affected heavily. As a result, the enclave's GDP shrank sharply during the second half of 2008 and early 2009 (Yellen, 2010).

The most common explanation offered for the severe impact of the recession on the export sector is that China's economy and in particular the manufacturing sector suffered from excess capacity (Lardy, 2012). Excess capacity, it was argued, was due to the high proportion of GDP that had been devoted to industrial and property investment, leading to the previously mentioned structural imbalances in the economy (Yongding, 2010). In 2007, for instance, the combination of fixed asset investment (FAI) and net exports to growth exceeded 60 per cent and could be considered the driving force in economic growth. Moreover, FAI growth had been for some time higher than that of GDP and, therefore, the obvious outcome is overcapacity growth at a rate greater than that of aggregate demand. Therefore, the case is that, if other parts of the economy cannot raise their level of demand, existing excess capacity can be absorbed only by yet further growth in FAI, which creates a feedback loop to increasing cycles in capacity growth. Eventually, however, FAI expansion will reach a peak beyond which it cannot go as a result of environmental, social, resource and other constraints, and the process will break down (Yongding, 2010).

Lardy (2012), however, offers a different perspective on the argument, claiming that for some products there will always be a degree of overcapacity if only on a temporary basis. The question is whether the level of alleged excess capacity is so large that it could have a deflationary impact, thereby affecting output, prices, profitability and ability to service loans, thereby leading to plant closures and unemployment. In discussing China, however, several factors need to be taken into consideration. The first is that historically Chinese firms have long tended to hold on to out-dated equipment rather than scrap it so that, if there was an upswing in demand, it could be brought back into use. Therefore, the importance of excess capacity in Chinese industry may have been overstated. Second, in mature economies growing at only 2 or 3 per cent per annum it is expensive to retain old equipment. This is less so in China where, under conditions of near continuous high growth, excess capacity may be reabsorbed quicker than it would in slower-growing ones, and so the retention costs are much lower. China's steel industry is often held as an example of the latter (Lardy, 2012; Lardy and Borst, 2013). This argument cannot be settled here, but there seems little doubt that falling exports did reveal excess capacity in particular industries, and the near inevitable outcome was firm closures and unemployment in export-oriented industries (Zhang, 2009).

That the recession led to rising unemployment is well documented but, because of difficulties with the official unemployment data resulting from flaws in data collection, a lack of national standardisation and deliberate distortion of the figures for political reasons, it is hard to arrive at an accurate estimate of the recession on China's unemployment rate (Duckett and Hussain, 2008). Moreover, often the data reflect only those workers who are eligible for inclusion in the unemployment registers and who live in cities (Schucher, 2009). Despite such hindrances, the registered urban jobless dropped between 2002 and 2007, when it reached 4 per cent. Subsequently the unemployment trend moved upwards, reaching 4.6 per cent in September 2009. By the following December, approximately 8.8 million residents featured on the urban registers as being unemployed, representing a rise of 560,000 in the last quarter of the year. Finally, it has been estimated that, as a consequence of the recession, some 20–36 million registered and unskilled migrant workers were rendered unemployed by the recession (Fardoust *et al.*, 2012; Zhang, 2009).

In addition to its severe impact on migrant workers, the crisis had an adverse effect on university graduates seeking employment. From 1999 onwards the government had continuously expanded higher education opportunities at a rate of circa 20 per cent per annum both to improve educational standards and relieve pressure on the graduate labour market. By 2008 the number of newly enrolled students had reached over 6 million. In that year graduate recruitment virtually dried up as most FIEs and banks, for example, stopped recruiting, to the extent that between 2 and 6 million graduates helped to swell the ranks of the unemployed (Zhang, 2009; Laike and Huzenga, 2010; Zhang *et al.*, 2010).

As Zhang (2009) has argued, the rising unemployment was considered by the government as a potential cause of social protest and unrest, and so the government continued to prop up failing firms by offering tax rebates on exports, intervening in the foreign exchange markets to inhibit any appreciation of the renminbi against the American dollar and reducing environmental and pollution controls in the affected areas to maintain employment. However, the vagaries of International trade are beyond the control of an individual government and so governmental attempts to maintain export volumes proved limited in an atmosphere of growing protectionist sentiment (Zhang, 2009).

China's response to the global crisis

In the autumn of 2008 China's rate of economic growth came to a near sudden halt by Chinese standards with the collapse of the export market, which can best be described as a major external economic shock, worsened by the demise of Lehman Brothers. The most obvious illustration of this was the collapse of the steel industry, where the fall in exports between August and September accounted for 54 per cent of the total fall in steel output (Yongding, 2009). The question that arises is, how did the Chinese government respond to the situation? The simple answer is that, like its counterparts in Western economies, the Chinese had little option but to utilise a mixture of monetary and fiscal policies as well as administering a somewhat old-fashioned economic stimulus to pump-prime activity. As Naughton (2009) pointed out, forcefully, such swift action, encapsulated in increasing government demand in the economy, prevented economic collapse and allowed growth to continue, albeit at a reduced rate initially, arguing that the Chinese response to the crisis was a classic Keynesian initiative that worked in textbook fashion (Naughton, 2009).

Though China had enjoyed almost 30 years of near-consistent growth rates in the three decades following the inauguration of the Open Door policy, progress was by no means linear and fluctuated between periods of inflation and deflation, with government policy being adjusted to deal with the appropriate situation. In 2004, with signs of overcapacity emerging, the government tried to dampen down new investment, especially in the steel industry, but in vain as new mills continued to be constructed. Generally the desire to invest continued and, in 2007, and coupled with strong export demand, inflationary pressure built up, leaving the authorities with little option but to tighten monetary policy and so prevent another property bubble arising (Yongding, 2009; Lardy, 2012). Circumstances then changed.

When the economic crisis arrived in the following year and its consequences began to bite deeply into both economy and society, there was little alternative open to the government but to throw deflationary policy levers into reverse and ease monetary policy to offset the economic consequences of the slowdown in world trade and its subsequent deflationary impact on China's GDP (Yongding, 2009).

Putting it simply, China's response to the situation was early, well-crafted and decisive. This was because it was in a strong financial position, mainly because of its high level of reserves which could support a substantial fiscal and credit expansion particularly in the light of a low budget deficit of below 3 per cent (Li *et al.*, 2012). There were those who decried the situation, arguing that the crisis was a demonstration that the entire reforming process had been a failure, that globalisation and the introduction of market forces had been detrimental to Chinese society as a whole and that the economy should return to the Marxist-Leninist model that had prevailed before the Open Door policy. Such conservative voices, though vocal, made little headway, and the government proceeded with its intended policy (Shusong, 2010).

At this point it is more than useful to look at what the Chinese authorities actually did to deal with the crisis. A looser monetary policy was followed initially. First, the government abolished the previous lending quotas that had been instituted to control the amount of bank lending (Lijun and Bo, 2010). Second, to ensure that the banks still had sufficient funds to meet their customers' legitimate demands, the government reduced the share of deposits that were normally placed with the central banks. There was no attempt to force the banks to increase their lending in early 2009. It was in their own interests to do so as they could earn higher rates of interest through lending than by placing their funds with the central bank or in lending them in the inter-bank market (Shusong, 2010; Lardy, 2010). In essence this was a step to maintain a necessary amount of liquidity and loanable funds in the economy.

Third, the government also took action to increase the demand for funds by lowering the benchmark bank rates used to guide the rates charged by banks on various maturities such as five-year loans, which were reduced from 7.7 per cent in September 2008 to 5.7 per cent by the close of the year. Additionally, deep cuts were made in mortgage rates for individual home buyers with a five-year loan reduced by as much as 18 per cent per month; for property investors, the 40 per cent minimum down payment on a mortgage was cut from 40 per cent to 20 per cent, and the compulsory penalty interest rate that applied to such investors which had been running at 1.1 times the Central Bank's benchmark rate was abolished. Similarly, the time period that investors had to hold on to property to avoid a subsequent sales tax was cut from five years to two years. Such pump-priming in the property market no doubt stimulated bank lending on a significant scale in the first half of 2009, when domestic loans rose by RMB 7.4 trillion, three times the level of increase in the first six months of the previous year. Thus, there is little doubt that mortgage lending played a more than significant role in helping to resuscitate China's economy (Lardy, 2010, 2012; Lardy and Borst, 2013).

Accompanying the recovery in property development was a straightforward financial stimulus that was injected into the economy, amounting to 4 trillion yuan, which was targeted at direct expenditure, as shown in Table 8.1. This funding was geared to real estate sector programmes and in total was considerably larger than the financial stimulus package of US $787 billion offered by President Obama in the United States. Moreover, a further difference between the American and Chinese packages was that, while the former was used primarily to pay off or reduce indebtedness, the Chinese version was designed to boost both spending and employment. Targeted primarily at infrastructural investment programmes, the package can be described as being of a Keynesian interventionist nature. The major part of the package was aimed at transport projects, which accounted for 37 per cent of total spending. Entailed in this was the intention of building 50 new airports and expanding another 90 existing ones at a cost of RMB 400 billion. A further RMB 600 billion was earmarked for upgrading the railway system through the creation of new lines. Additionally, RMB 1 trillion was designated for road, seaway and local transit systems (Schuller and Schuller-Zhou, 2009). What

Table 8.1 Spending structure of 4 trillion yuan stimulus package (in billion yuan)

Housing construction for low-income households	280
Spending on rural infrastructure, housing and incomes	370
Investment in transportation	1,800
Investment in medical services, culture and education	40
Spending on ecological protection	350
Technical innovation and economic restructuring	160
Sichuan post-earthquake reconstruction	1,000
Total	4,000

Notes: 1 This table is based on a press conference release by P. Zhang, Head of China's National Development and Reform Commission, 27 November 2008

2 Renminbi is the official name of China's currency. The Yuan is the name of a unit of the Renminbi. It is similar to the relationship between Sterling and the British pound.

Sources: Yongding (2009); Liu (2009)

is important is that many of these projects were labour intensive, with each of the new high-speed railway tracks alone employing upwards of 10,000 workers. Finally, about 25 per cent was devoted to post-earthquake reconstruction in Wenchuan County in Sichuan Province, where the damage had been extensive (Yongding, 2009).

Lastly, the government inaugurated industrial revitalisation programmes to improve the competitiveness of the ten 'pillar industries' that had been identified during the 1980s and 1990s as forging the future of Chinese industry in the global economy: the automotive industry, steel, shipbuilding, textiles and clothing, machinery, electronics and information technology, light industries, petrochemicals, non-ferrous metals and logistics. The new support for these included tax incentives, industrial subsidies, privileged access to government contracts and procurement, funding for technology upgrading, overseas investment promotion and much more concentration on promoting domestic brands. In the case of the latter, Chinese consumers often preferred to buy foreign imported goods rather than their domestically produced counterparts simply because of their superior brand recognition attributes and perhaps simply to show that they could afford the higher price of imports (Donnelly *et al.*, 2010; Yang and Huizenga, 2010).

Shortly after the central government introduced its stimulus package, many local, provincial and urban governments brought out their own revival plans which were estimated to reach around RMB 18 trillion (Yang and Huizenga, 2010). Of particular importance in both programmes was the desire to improve living standards in the rural economy, as much of China's wealth is concentrated in the major cities in the Eastern and Southern seaboards, which only adds to the country's social imbalances especially in Central and Western China. One way of achieving this was through the rural appliance programme to stimulate consumption among low-earning rural families. In December 2008 China's State Council issued the 'Opinion on Stimulating Circulation and Expanding Consumption' that included 20 measures to boost consumption by offering a discount of 13 per cent on the purchase price of household appliances such as refrigerators, televisions, air-conditioning units and washing machines up to a specific sum. It has been argued that this had a positive impact in rural society by encouraging destocking of goods held by firms, offsetting the decline in export demand to a limited degree and helping to improve the quality of life among the disadvantaged, even if the overall impact in raising consumption was probably at best marginal (Fardoust *et al.*, 2012). Finally, in the increased spending on rural health care and on urban social welfare, the minimum livelihood guarantee (*dibao*) was also an important part of the package for those on low incomes. The former reduced the need for low-income rural families to set aside so much of their wages to pay for medical services, while the latter eased the same burden for many poor urban households, thereby increasing their immediate disposable income and the need for precautionary savings (Bank of England, 2011; Fardoust *et al.*, 2012).

Space precludes further discussion of the nature of both stimuli, and so the question arises of how the package was financed. Essentially there were four clear avenues pursued. First, the central government was responsible for 25 per cent in the form of grants, subsidies in interest rates and direct injections of capital into government-originated projects. Second, budget deficit funding was followed through issuing of government bonds, which was relatively easy as the ratio of budget deficit to GDP was only 0.6 per cent in 2008 and calculated not to exceed 3 per cent the following year. Third, the government issued bonds to the value of RMB 200 billion on behalf of local, urban and provincial governments to plug the gaps in funding local projects. Fourth, bank lending was the prime source of funding for local authorities (Yang and Huizenga, 2010).

Attractive though the package was, it was not without its critics. China's recovery was relatively swift in comparison with that of other countries, notably the United States, the EU and Japan, even if GDP growth in 2009 was below that achieved in earlier years. The key criticism was that China's recovery to a high growth rate was unsustainable since it was predicated on what appeared to be a one-off stimulus, financed primarily via various forms of bank lending (Roach, 2009). This, it was postulated, led to the creation of asset bubbles in property and equity markets as funds originally earmarked for investment purposes leaked into these sectors. A second and perhaps justifiable criticism was that industrial investment simply exacerbated the existing problem of excess capacity which ultimately would depress both prices and profits. In turn, this might affect firms' ability to service debt and so lead to a rise in non-performing loans (Lo, 2010; European Chamber, 2009). A third charge is that the sheer scale of the stimulus might undermine China's strong fiscal position. Indeed, the deficit did not even top 2 per cent in 2009, a mere fraction of the scale endured in other advanced industrial nations: China's outstanding debt in sum remained at 20 per cent of GDP. However, critics allege that this masks a massive hidden government debt. Lastly, Pettis (2010) claims that the entire package simply served to increase the already serious imbalances in the economy, namely industrial investment, exports and property, to the neglect of consumption. One further niggling caveat is that a number of the projects enacted by subnational authorities might well have been already prepared 'shovel ready', i.e. these may have been 'off-the-shelf' favourites of specific officials who seized the opportunity to implement them even if the projected rates of return were dubious (Yonding, 2010).

While there is more than a grain of truth in the above criticisms, Lardy (2012) is quite dismissive of them. In dealing with the property bubble, he points out that, when managing the growth of lending, the authorities, as early as mid-2009, took action to limit lending, even if it did not spike until January the following year. Then mandatory lending quotas on individual banks were reinstated, as was the amount that could be lent over the first two quarters of that year. Reserve ratios were also raised by 50 points to reduce excess reserves and to serve as an indicator that monetary policy was tightening. For example, the China Bank Regulating Commission announced that no longer would banks be permitted to count subordinated debt and hybrid capital as part of their tier two capital. Such moves and others were specifically designed to restrain excess bank lending (Lardy, 2010, 2012).

In dealing specifically with mortgages, there was never any intention of allowing the market to grow unchecked, and so the Chinese government intervened in the market just as had happened in 2007 when there were similar signs of overheating. In December 2009 the 40 per cent down payment for loans to property investors was reinstated and the resale tax (which was time limited) lengthened from two years to five years once more. There were clear disincentives to head off property speculation and cool the market. The outcome was that property sales dropped quickly in late 2009 and in 2010. What also helped to moderate sales to individuals as distinct from investors is that Chinese households are much less leveraged than their counterparts in other countries, and so the share of debt to house purchase is relatively small. This is a reflection of the high level of down payments required when buying a house in China and also perhaps of a propensity for buying a property with cash. Therefore, few Chinese find themselves in negative equity when prices fall, and so moderation did not present any threat to financial institutions as in the United States (Lardy, 2012).

Was excess capacity in industry and property created? Superficially this seems an attractive proposition but, as argued earlier in the paper, the absorption of excess capacity is relatively easy in a rapidly expanding economy such as China's compared to less rapidly growing, mature, advanced countries. If there was such capacity created, then it is likely to have been temporary in nature. The key issue, though, is not excess capacity *per se*, but whether the new capacity was so widespread that it could lead to downwards pressure in prices to the extent that it could impact adversely on the banking system through default on payments. Such an outcome does not seem to have resulted as a consequence of the one-off 2008 stimulus, with the possible exception of the steel industry, where in 2009 there was an excess capacity of 15–20 per cent, which is greater than the entire steel output of the United States and Japan (European Chamber, 2009). Even this, however, really pays insufficient attention to the rate at which an expansion in China's growth rate can increase the pace of absorption of excess capacity in steel production. Lastly, it needs recalling that 50 per cent of the investment boom of 2009 was targeted primarily at infrastructural projects, 13 per cent to leasing and 10 per cent to property, rather than being pointed at traditional industries, which helped to curtail the opportunities to create excess capacity (Lardy, 2012; Yongding, 2009).

The third charge levied against the stimulus is that of increasing government debt. Most of the stimulus was funded by credit rather than deficit spending and so, therefore, government debt remained low. However, a significant amount of the medium-to-long-term bank lending for infrastructure improvements was directed to quasi-governmental agencies, often referred to as platform companies. Under Chinese law, local governments are not permitted to borrow or indulge in deficit financing, but lending to such agencies is legal. Critics argue that it is unlikely that many such companies will be able to repay loans and so the burden will fall on local governments, many of which had agreed to act as guarantors. The question arising is just how large will such debts be and will they be repaid? The scale of loans is estimated to be around RMB 5–6 trillion, equal to 20 per cent of China's GDP. By any standard this figure is huge; and it is probably fair to say that it is unlikely that all of them will be honoured in repayment, and that in future some of these might become non-performing loans. On the positive side, the infrastructural projects may bear fruit in contributing to economic growth, increasing tax revenues and increasing land values that could be either sold or leased for development to realise a return (Lardy, 2010).

A final charge levelled against the stimulus was that it paid too much attention to investment and too little to consumption and so retarded attempts to correct the internal imbalances within the Chinese economy. However, in 2009 consumption growth was relatively strong, outstripping the growth of GDP for the first time since 2000. There were several factors that explain the strength of consumption. First, booming investment helped to offset at least some of the job losses in export industries. In total, some 11 million jobs were created in 2009. Second, transfer payments were increased to the elderly enterprise retirees by 10 per cent and by a third to those in the poorest households, with the latter amounting to some 70 million people (Lardy, 2012). Third, the authorities tried to stimulate consumption by tax reduction in industries such as automobiles to boost the market, especially at the small car end. In rural areas a 'scrappage' scheme was introduced (Tan, 2012). Overall, car sales rose by around 50 per cent nationally; this was accompanied by increasing sales of consumer durables, especially in rural areas. Finally, rising consumption was reflected in increased household borrowing. RMB 4.1 billion was borrowed by consumers to finance house purchasing, with a further RMB 1.1 billion being used to finance more general consumption. Overall the indications are that consumption in 2009 rose by around 10 per cent, well ahead of the growth rate in GDP (Lardy, 2010, 2012).

Though it is hard to arrive at a firm verdict on the success or otherwise of the policies pursued by China in response to the effects of the global recession, it is safe to say that it was spared the same recessional effects that fell upon the United States and Europe (Yang and Huizenga, 2010). There is no doubt that policy effects were positive in maintaining macroeconomic stability and in preventing a total collapse in exports after the nadir of March 2009. Though GDP growth rates did fall, by international standards they remained good, at circa 9 per cent in 2008 and circa 8 per cent the following year, reaching 11.4 per cent in the second quarter of 2009. Additionally, imports, especially of crude oil, continued to stay at a high level, rising by 55 per cent over the same period to meet a revival in burgeoning demand in December 2009 (Lardy, 2012; Lardy and Borst, 2013).

While the above paragraphs describe the policy outcome, albeit very briefly, they fail to offer an explanation of why China's approach to the recession worked. One explanation proffered is that success was down to the sheer scale and strength of the total package, which allowed a series of multipliers to make a positive impact on the economy. First, policy was counter-cyclical and in contrast to the pro-cyclical policies normally pursued in emerging conomies, especially at subnational levels, and in advanced countries such as the United States, where under the federal system of government individual states were obliged to balance their budgets through austerity policies even in the face of President Obama's attempts to revitalise the economy. Moreover, as elsewhere, public sector employment looms large in unitary states such as China, with 80 per cent of total state expenditure being carried out at subnational levels. Therefore, increased central government spending to increase demand and liquidity was bound to have a positive effect on employment levels as it percolated down through the provincial and urban systems via both the public and private sectors, making the effect of direct government intervention particularly strong (Fardoust *et al.*, 2012).

Second, China was in a strong position to finance such a large stimulus, and its high accumulation of foreign reserves provided solid support for the stimulus without any adverse effect on the value of the renminbi. Additionally, multipliers tend to be larger and perhaps more effective when the level of government debt is low. Moreover, as said before, policy implementation was swift, and clear targets were identified; and perhaps this enabled investment levels to remain relatively buoyant at a time when they were falling in other countries. In essence the stimulus prevented the 'onset of the vicious cycle of lower investment, lower consumption and lower profitability' (Fardoust *et al.*, 2012).

A final question is, to what extent did the almost emergency policy followed by the Chinese state affect the restructuring needs that had been identified before the crisis from being addressed? The prevalent view is that, if China is to continue to advance and escape what is called the 'middle income effect' on its economy, further action must be taken. Lardy (2012) argued that the 2008–2009 stimulus did not address the appropriate issues; but perhaps being a one-off crisis measure, it was never intended to do so – being targeted at one specific instance, which may have reinforced the triangle of investment, exports and property in the short term. However, there is little doubt that, in the medium-to-long term, the Chinese economy will benefit from a new economic model which will require major reforms if it is to move forward from being a developing economy to a modern advanced entity with satisfactory rates of growth in GDP and able to compete fully on the international stage (Liu, 2010). The major change advocated is to switch from an overemphasis on investment and exports to consumption. Effecting such changes will not be easy and will take time within a context of prudent but proactive economic management (Lardy, 2012; Feurberg, 2012; Yueh, 2012).

This topic demands a book in its own right. All one can do here is to outline briefly without going into detail a number of steps that might be taken to rebalance the economy. What is required, on the one hand, is a relative decline in industrial investment levels, a lesser dependence upon export growth and a containment of property development. Counterbalancing this is an increased emphasis on the growth of consumption, an increased growth in labour-intensive service industries and a reduction in regional and income inequalities (Lardy and Borst, 2013; Yueh, 2012; Wang, 2010).

Clearly the implied scale of action is vast; but similar rebalancing acts have been carried out in the past, notably in Taiwan and Japan. Entailed in these changes are the need to remove the market distortions in finance and banking by liberalising interest rates, exchange rates and energy prices. Similarly, the renminbi needs to be allowed to appreciate to an appropriate level. This would make exports more expensive, but reduce the price of imports, which in turn could stimulate consumption and possibly reduce China's external surplus as well as improve trade imbalances in the world economy, especially those with the United States. Additional steps could be an increase in outwards foreign direct investment (OFDI) and perhaps China's playing a wider role in the international financial system via the International Monetary Fund (IMF) (Zhang *et al.*, 2010). Further improvements could be made by reining back on controls on energy and gasoline prices, which prevent industry and consumers from paying full costs for the product: this leads to energy companies' refining units racking up losses when prices are high. Such price distortions are also found in electricity generation as a result of fluctuations in coal prices. The effect of this is to indirectly subsidise intensive energy users in manufacturing industry, which means that these industries can be more profitable than those in the services sector. Until such distortions and inequalities are corrected, the imbalance between services and manufacturing will persist (Lardy and Borst, 2013).

A final imbalance is that of income inequality, which if modified would undoubtedly lead to an increase in consumption. China spends only 5.7 per cent of its GDP on health and welfare. Required here are improvements in the social and welfare safety net through increases in transfer payments to the rural poor and urban disadvantaged, which would improve their lifestyle by reducing the precautionary amounts saved for medical costs and, particularly, primary school fees. Furthermore, if targeted accurately, such transfers in pensions, etc. would also help to narrow the differences in incomes between those workers in the eastern coastal regions and those living in western and central regions. Additionally, taxation could be made more progressive and firms could be forced to make higher contributions towards employees' pension and health funds. While such measures in themselves may not necessarily increase household consumption *per se*, government expenditure in these areas through improved medical, health and educational provision could have quite a dramatic impact on consumption through both the public and private sectors. Lastly, reforming the *hukou* registration system might ease the plight of migrant workers in cities and help in improving the quality of life for them and their families (Lardy and Borst, 2013).

In essence, getting rid of price distortions, reducing inequalities and boosting consumption will take time and a considerable amount of reforming zeal on the part of the central government to enforce its will against what might be described as the vested interests of the more conservative elements in the ruling party and in provincial governments. To achieve these goals it will probably be essential to reduce the share of industrial investment in GDP to around 38 per cent, with the balance tilting towards services accordingly. The reason for investment remaining in the 'high 30s' percentage-point range is simply that China's high rate of urbanisation will continue for the foreseeable future, which will demand ongoing investment in infrastructure, without the emergence of housing bubbles. The urban population takes up approximately 45 per cent of the total population at present, but is expected to expand to 75 per cent in the next 30 years. This means that some 400 million people will move to the cities, resulting in the equivalent of another 400 cities with in excess of a million people (Weiying, 2010). It's in this context that it needs to be remembered that the overall target aimed at is a level of growth that is sustainable, which means that China may have to settle for single-digit rather than double-digit GDP growth for a decade or perhaps longer (Jinglian, 2010).

Conclusion

Several conclusions or lessons can be drawn from China's experience in the recent global economic crisis. The first of these is that the crisis itself helped expose just how much China has become integrated into the global economy since the onset of the Open Door policy in 1978. China's economy was not immune from external shocks, occasioned this time by the slowdown in growth in the G3 countries. Second, China's economic overdependence on industrial investment, property development and exports was amply demonstrated. Third, the latter revealed internal imbalances in China's economic structure: 'unsteady, imbalanced, uncoordinated and unsustainable' (Wen Jibao, 2007). In other words, the economic model followed was flawed and needed changing.

In looking at how the crisis impacted on China it can be concluded that the pattern differed from that in the United States, Japan and the EU. It was China's tradeable economy that suffered through a sudden fall in exports, where in the other areas the consequences of the crisis came primarily through the financial sector. Nevertheless, all four areas suffered an accompanying rise in unemployment, and economic growth faltered.

China's response to the crisis was somewhat unique in that, rather than follow the austerity route, as in parts of Europe including the UK, a near-Keynesian stimulus policy was introduced and monetary policy was relaxed to prevent the economy slipping further. Given the fall in GDP and the subsequent effect of the stimulus, which was essentially a short-term measure, the outcome was that the economy turned upwards in just over one year and unemployment began to diminish, even if the injection of capital into the economy did reinforce imbalances: it is safe to conclude that in the short term the stimulus did achieve its objectives.

The problems of economic and social imbalances in China remain, and further action is required to reduce the distortions in the financial system, as well as in regional and social inequalities. It remains to be seen, however, whether or not the current political and institutional structures in China are capable of further reform to drive society forward.

References

Angang, Hu (2010) *China's Experience in Coping with the International Financial Crisis*, Economists' Discussion on China's Economy, Foreign Languages Press, Beijing.

Bank of England (2011) *China's Changing Growth Pattern*, Bank of England Research and Analysis, Bank of England, London.

Cai, F., Wang, F. and Yang, D. (2002) 'Regional Disparity and Economic Growth in China: the impact of labour market distortions', *China Economic Review*, 111: 1–16.

Donnelly T., Begley, J. and Collis, C. (2010) 'Towards Sustainable Growth in the Chinese Automotive Industry: internal and external obstacles and comparative lessons', *International Journal of Automotive Technology and Management*, Vol. 10, No. 2/3: 289–304.

Duckett, J. and Hussain, A. (2008) 'Tackling Unemployment in China: state capacity and governance issues', *The Pacific Review*, Vol. 21, No. 2, May: 211–229.

Elwell, C. (2013) 'Economic Recovery: sustaining US economic growth in the post-crisis-economy', *Congressional Research Service*, CRS Report for Congress, 7-5700, R 41332, Washington, DC, pp. 1–27.

European Chamber (2009) *Overcapacity in China: causes, impacts and recommendations*, November, available at www.europeanchamber.com last accessed 3 May 2014.

Fardoust, S., Lin, J. and Luo, X. (2012) 'Demystifying China's Fiscal Stimulus', Policy Research Working Paper, No. 6221, *The World Bank*, Development Vice Presidency, Office of the Chief Economist, Washington, DC, October.

Feurberg, G. (February, 2012) 'Chinese Economic Growth Requires Restructuring Economy', *The Epoch Times*, available at www.theepochtimes.com/n2/china-news/chinese-economic-growth-requires-restructuring-economy-192108.html last accessed 2 June 2014.

Guthrie, D. (2009) *China and Globalisation: the social, economic and political transformation of Chinese society*, Routledge, New York.

Jinglian, W. (2010) *How to Change the Development Mode?* Economists' Discussion on China's Economy, Foreign Languages Press, Beijing.

Hopkins, K. (2014) 'China Set To Oust US as World's No. 1', *The Times*, 1 April, p. 44.

Laike, Y. and Huizenga, C. (2010) 'China's Economy in the Global Economic Crisis', in *The Financial Crisis of 2008–09*, United Nations, New York.

Lardy, N. (2010) 'The Sustainability of China's Recovery from the Global Recession', *Policy Brief*, Petersen Institute for international Economics, Washington, DC.

Lardy, N. (2012) *Sustaining China's Economic Growth After the Global Financial Crisis*, Petersen Institute for international Economics, Washington, DC.

Lardy, N. and Borst, N. (2013) 'A Blueprint for Rebalancing the Chinese Economy', *Policy Brief*, Petersen Institute for international Economics, Washington, DC.

Li, L., Willett, T. and Nan, Z. (2012) 'The Effects of the Global Financial Crisis on China's Financial Markets and Macroeconomy', *Economics Research International* Vo. 12, No. 11, 961694, available at http://dx.org/10.1155/2012/961694 last accessed 13 January 2014.

Lijun, M. and Bo, W. (2010) 'Lending Caps To Reduce Liquidity', China Daily, 21 January, p. 10.

Liu, L. (2009) 'Impact of the Global Financial Crisis on China: empirical evidence and policy implications', *China and World Economy*, Vol. 17, No. 6: 1–23.

Liu, S. (2010) *A Turning Point: China's economy after the bubble*, Economists' Roundtable Discussion on China's Economy, Foreign Languages Press, Beijing.

Lo, B., (2010) *China and the Global Financial Crisis*, Centre for European Reform, London.

Morrison, W. (2009) *China and the Global Financial Crisis: implications for the United States*, Congressional Research Services, 7--5700, Washington, DC.

Nan, L. (2009) 'China's Policy Responses to the Global Financial Crisis: efficacy and risks', Conference on *Global Financial Governance: challenges and regional responses*, Berlin 2–4 September.

Naughton, B. (1995) *Growing out of the Plan: Chinese economic reform 1978–1993*, Cambridge University Press, New York.

Naughton, B., (2009) 'Loans, Firms and Steel. Is the State advancing at the expense of the private sector?', *China Leadership, Monitor*, No. 30.

Pettis, M. (2010) 'China Has Been Led by Bears and Bulls Alike' *Financial Times*, 26 February, p. 11.

Rawski, T. (1994) 'Progress without Privatisation: the reform of China's State industries', in *Changing Political Economies: privatisation in post-communist and reforming communist states*, pp. 27–52, Milor, V. (ed.), Lynn Reinner, Boulder, CO.

Roach, S. (2009) 'An Unbalanced World Is Again Compounding Its Imbalances', *Financial Times*, 7 October, p. 23.

Schucher, G. (2009) 'China's Unemployment Crisis: a stimulus for change?', *Journal of Current Chinese Affairs*, Vol. 38, No. 2: 121–144.

Schuller, M. and Schuller-Zhou, Y. (2009) 'China's Economic Policy in the Time of the Global Financial Crisis: Which way out?', *Journal of Current Chinese Affairs*, Vol. 38, No. 3: 165–181.

Shusong, B. (2010) *China's Economic Recovery Management in the 'Post-Crisis period'*, Economists' Roundtable Discussion on China's economy, Foreign Languages Press, Beijing.

Tan, Z. (2012) *The Development of the Chinese Car Industry*, unpublished PhD thesis, SURGE, Coventry University, UK.

The Times (2014) 'China's Crossroads', 1 April, p. 30.

Wang, Z. (2010) *How Will China Shift its Economic Development Mode?*, Economists' Roundtable Discussion on China's Economy, Foreign Languages Press, Beijing.

Weiying, Z. (2010) *How Big is China's Market Potential?*, Economists' Roundtable Discussion on China's Economy, Foreign Languages Press, Beijing.

Wen, Jibao (2007) 'The Conditions for China To Be Able To Continue and Guarantee Stable and Rapid Economic Development', press release, 16 March, available at www.npc.gov.cn last accessed, 5 March 2014.

Williams, J. (2012) 'The Federal Reserve and Economic Recovery', Federal Letter of The Reserve Bank of San Francisco, CA, 17 January.

Xiaonian, X. (2010) *What Happened to China and America Before and After the Financial Crisis?*, Economists' Roundtable Discussion on China's Economy, Foreign Languages Press, Beijing.

Yang, l. and Huizenga, Y. (2010) *China's Economy in the Global Economic Crisis: impact and policy responses*, UNCTAD publications, New York.

Yellen, J. (2010) 'Hong Kong and China and the Global Recession', *FRBSF Economic Letter*, Federal Reserve Bank of San Francisco, CA, 8 February.

Yongdin, Y. (2009) 'China's Policy Responses to the Global Economic Crisis and Its Perspective on the Reform of the International Monetary System', paper prepared for the AEEF held at Kiel Institution of the World Economy, Kiel, Germany, 7 July.

Yongding, Y. (2010) *The Impact of the Global Financial Crisis on the Chinese Economy and China's Policy Responses*, Third World Network, Jutaprint, Malaysia, available at www.twnside.org.sg/title2/ge/ge25.pdf last accessed 2 June 2014.

Yueh, L. (2012) China's Strategy Towards the Financial Crisis and Economic Reform, London School of Economics, available at http://eprints.lse.ac.uk/44206/1/

__Libfile_repository_Content_LSE%20IDEAS_Special%20Reports_SR012%20
China%27s%20Geoeconomic%20Strategy_China%27s%20Geoeconomic%20
Strategy%20_China%E2%80%99s%20Strategy%20towards%20the%20Financial%20
crisis%20and%20Economic%20Reform%20%28LSE%20RO%29.pdf last accessed 2
June 2014.

Zhang, M. (2009), *The Impact of the Global Crisis on China and Its Reaction*, Elcano
Royal Institute Analyses, Elcano Royal Institute of International and Strategic Studies,
Madrid, available at http://isn.ethz.ch/Digital-Library/Publications/Detail/?ots591=
0c54e3b3-1e9c-be1e-2c24-a6a8c7060233&lng=en&id=146353 last accessed 2 June
2014.

Zhang, Z., Wei, L. and Nan, S. (2010) 'Handling the Global Financial Crisis: Chinese
Strategy and Policy Response', Working Paper, Business School, University of
Durham, UK.

Zhenzhong, W. (2010) *How Will China Shift Its Economic Development Model?*,
Economists' Roundtable Discussion on China's Economy, Foreign Languages Press,
Beijing.

9 North American and European responses to crisis in the motor vehicle industry

James M. Rubenstein

The severe global recession of 2008–2009 had a major impact on the world's motor vehicle industry. Worldwide production and sales declined sharply, and two of the ten largest carmakers filed for bankruptcy protection. This chapter compares the motor vehicle industries in Europe and North America during the severe recession and the policy responses in the two regions to these conditions.[1]

The chapter is divided into three sections. First is a comparison of motor vehicle industry trends in Europe and North America at the start of the severe recession in 2008. Second is a comparison of policy responses in Europe and North America to auto industry conditions during the recession. Third is a discussion of on-going challenges in the European and North American motor vehicle industries in the aftermath of the severe recession.

Industry conditions at the onset of the severe recession

The severe recession that started in 2008 brought to an end an extended period of growth in production and sales of motor vehicles. The collapse of credit markets in the early days of the recession played a major role in the severe decline in sales and production of vehicles. Most buyers of new vehicles finance the purchase through loans, which became difficult to obtain. Similarly, the vehicle manufacturers—both the parts suppliers and the assemblers—were also unable to obtain the credit needed to purchase materials and finance day-to-day operations.

During the recession, worldwide motor vehicle sales declined from 71.2 million in 2007 to 65.4 million in 2009. Production declined even more sharply, from 73.3 million in 2007 to 61.8 million in 2009, the severest global decline since World War II. As the recession deepened and lingered, production fell off more rapidly than sales, because carmakers struggled to reduce the inventory of unsold vehicles.

The world's major regions of vehicle production and sales were impacted differently by the recession. In North America, new vehicle sales declined 33 percent between 2007 and 2009, from 19.3 million to 12.9 million, whereas in Europe sales declined only 14 percent, from 18.9 million to 16.2 million vehicles. The recession did not dampen extremely high growth rates in most of Asia, where sales actually increased 20 percent between 2007 and 2009, from 23.6 million to

28.3 million vehicles, led by a 55 percent sales increase in China. The principal exception in Asia was Japan, where sales declined 13 percent during the two years. Similarly, between 2007 and 2009, production of vehicles declined 43 percent in North America, from 15.5 million to 8.8 million, and 22 percent in Europe, from 19.7 million to 15.3 million, yet increased 4 percent in Asia, from 30.7 million to 31.8 million, as growth in China more than offset a decline in Japan.[2]

With the onset of the recession in 2008, motor vehicle industry executives and government leaders in Europe and North America made differing assessments. In North America, the condition of the Detroit 3 (the three U.S.-headquartered carmakers, General Motors, Ford, and Chrysler) was regarded as a structural crisis. "By midsummer of 2008, the nightmare scenario was coming to life—soaring fuel prices, a miserable economy, no credit for consumers" (Klier and Rubenstein, 2012: 36; Vlasic, 2011: 284). As its financial condition deteriorated through 2008, GM held discussions about a possible merger with Chrysler and Ford and tried to raise funds by selling assets, borrowing, and applying for government assistance. None of these initiatives were successful (Vlasic, 2011: 286–287, 305). "By the beginning of December 2008, GM and Chrysler could no longer secure the credit they needed to conduct their day-to-day operations, according to Chrysler CEO Robert Nardelli" (Congressional Oversight Panel, 2011: 9). GM posted a near-record loss of $30 billion in 2008 and entered 2009 with a cash supply of only $14 billion. "General Motors had weeks—maybe days—before it defaulted on billions of dollars in payments to its suppliers" (Vlasic, 2011: 329). Privately held Chrysler, acquired by Cerberus Capital Management from DaimlerChrysler a year before the recession in 2007, also had a dangerously low supply of cash to meet day-to-day obligations.

Meanwhile in Europe, the motor vehicle industry was considered to be in a cyclical pattern. Historically, sharp short-term cyclical changes characterized Europe's motor vehicle industry. Double-digit annual percentage changes in sales and production within individual companies and countries were common as recently as the 1990s. Wars and economic conditions severely disrupted the industry on several occasions. The fortunes of individual companies waxed and waned. Sales would increase at the introduction of a new model, drift downward towards the end of the model's life, and increase again with the arrival of the next model. In contrast, between the early 1990s and the onset of the severe recession, annual sales figures changed much more modestly. After nearly two decades of modest annual changes, carmakers had come to believe that the historical pattern of cycles had been replaced by stability, with demand primarily for replacement of older vehicles. The downward drift of production and sales in Europe during 2008 was widely seen as a return to the long-term cyclical pattern and an end to a period of unusual stability.

For example, in October 2008, European Industry Commissioner Günter Verheugen suggested that the motor vehicle industry faced what he termed a "tough time," in part because of the global recession, but also because demand for new vehicles had "dampened" in anticipation of higher EU emission taxes. "Few people are sure what sorts of additional [emissions] taxes their purchases will be subject to" (*Spiegel On-line International*, 2008). In November 2008, at a Berlin conference sponsored by *Automobilwoche*, CEO Martin Winterkorn reported that "Volkswagen has held up well," and Ford of Europe executive Birgit Behrendt reported that her company was "performing relatively well." Supplier executives were said to be confident that "they will be OK" (*Automotive News Europe*, 2008). In December 2008, Sigrid de Vries, a spokesperson for the European Automobile Manufacturers' Association (ACEA), the region's principal trade association, told *The New York Times* that the European car market was what she termed "struggling" (Jolly, 2008).

Policy response in North America

In North America, government leaders concluded that the region's motor vehicle industry was in a structural crisis, requiring a bold response. The region's entire motor vehicle industry was threatened because two leading carmakers were about to run out of cash. Restructuring the two weakest carmakers would strengthen the entire North American motor vehicle industry. The initiative can be considered a rare example in North America of implementing an industrial policy and a regional policy, although these terms were not used by the region's leaders.

Regional policy in North America

The core regional policy component of the intervention in the motor vehicle industry was recognition that the liquidation of GM and Chrysler would have a disproportionately large impact on specific areas within North America. A regional policy would have an impact on the motor vehicle industry at two scales: at the international scale within North America and at the interregional scale within the United States.

At the interregional scale within the United States, implementation required overcoming irreconcilable differences between northern and southern states that made Congressional action impossible. Nearly all U.S. motor vehicle production had clustered in the late twentieth century in the interior of the country, in an area known as auto alley;[3] but within auto alley important differences developed between the northern and southern portions. The northern portion of auto alley contained 17 'Detroit 3' assembly plants and only one foreign-owned one, whereas the southern portion of auto alley contained 16 foreign-owned assembly plants and only two owned by the Detroit 3.

Congressional consideration of the plight of the Detroit 3 carmakers began with testimony by Chrysler CEO Robert Nardelli, Ford CEO Alan Mulally, GM CEO Rick Wagoner, and United Auto Workers President Ron Gettelfinger before the Senate Committee on Banking, Housing, and Urban Affairs on November 18, 2008 and before the House Committee on Financial Services the next day. Few committee members were convinced by the testimony to appropriate open-ended loans for the purpose of maintaining the carmakers' day-to-day operations. The three CEOs testified to the two committees again on December 4 and 5, 2008, this time providing more compelling evidence that they would soon run out of cash. The House of Representatives passed by a 237:170 vote legislation authorizing loans to the carmakers. Support for the bill was positively correlated with a high level of Detroit 3 employment in the Congressional district, and opposition was positively correlated with a high level of employment at foreign-owned carmakers (Nunnari 2011: 1). In the Senate, on December 11 a vote to end debate on the House-passed bill was 52 in favor and 35 against, but 60 votes were needed on the procedural motion. After considering other options, the Senate abandoned further action on the issue, and the bill died (Cooney *et al.*, 2009: 3–8).

Faced with Congressional inaction and the urgent pleas of both the Detroit 3 and international carmakers, President George W. Bush, in the final month of his administration, and President Barack Obama, in the first month of his administration, both took forceful executive action to aid Chrysler and GM. President Bush issued an executive order on December 19, 2008, permitting the Treasury Department to establish the Automotive Industry Financing Program (AIFP) with funds from the Troubled Asset Relief Program (TARP) under the *Emergency Economic Stabilization Act* (EESA) of 2008 (Cooney *et al.*, 2009: 8). TARP loans made it possible for Chrysler and GM to stay afloat during the transition from the Bush Administration to the Obama Administration (Cooney *et al.*, 2009: ii). President Obama on February 16, 2009, appointed a Presidential Task Force on the Auto Industry to devise a strategy for dealing with Chrysler and GM. Treasury Secretary Timothy Geithner and National Economic Council Director Lawrence Summers were co-chairs, and Steven Rattner, co-founder of the hedge fund Quadrangle Group, was lead advisor. According to IHS Automotive financial analyst Aarom Bragman, "the Bush administration ... acted to save [Chrysler and GM] from an uncontrolled bankruptcy and shutdown. The Obama administration's role was to fix them" (Greenberg, 2013).

Through the Bush Administration's TARP commitments, Chrysler and GM each received a \$4 billion loan on December 29, 2008. GM received a further \$5.4 billion on January 16, 2009, and \$4 billion on February 17, 2009, as part of the Bush Administration's December 2008 commitments. The Obama Administration committed another \$36.8 billion to GM, bringing the total TARP commitment to \$50.2 billion. Chrysler received another \$6.9 billion from the Obama Administration, bringing the total TARP commitment to \$10.9 billion. General Motors Acceptance Corporation (GMAC), GM's credit affiliate, received \$17.2 billion in TARP loans, including \$6 billion from the Bush Administration and \$11.2 billion from the Obama Administration. The Bush Administration also provided Chrysler Financial with \$1.5 billion in TARP loans (Canis and Webel, 2011: 6, Table 2).[4]

At the international scale within North America, implementation required cooperation among the governments of the United States, Canada, and Ontario on a coordinated multijurisdictional response. The governments of Canada and Ontario were clearly junior partners with the U.S. government. The troubled carmakers were U.S.-based, and the U.S. government could command far more resources to undertake financial restructuring of troubled carmakers than could its neighbors to the north. Nonetheless, Canada invested its "fair share" in the auto industry restructuring, approximately 10 percent, roughly equivalent to its share of the two countries' population. Approximately two-thirds of that 10 percent came from the national government and one-third from Ontario. Ontario is the leading governmental jurisdiction within North America in number of vehicles assembled and number of motor vehicle parts plants. In view of the importance of its motor vehicle production, Ontario has a ministerial-level office to keep close watch on industry conditions. As a result, it was in a position to provide critical expertise to the U.S. Department of Treasury in setting up and implementing the details of the restructuring programs.

Industrial policy

The essence of the industrial policy was fear that the demise of one or two of North America's large carmakers could threaten the viability of all carmakers. At first glance, government intervention to rescue troubled carmakers seemed counterintuitive. If, for example, Chrysler were liquidated, would not the remaining carmakers be strengthened by gaining its defunct competitors customers? The argument for liquidating at least one troubled carmaker was strengthened by the fact that the three U.S.-based carmakers had been losing market share for many years. The Detroit 3 had held 73 percent of the U.S. market in 1995, but their share declined to 65 percent in 2000, 56 percent in 2005, and 49 percent in 2008.

The Detroit 3 carmakers were especially vulnerable during the severe recession in part because a relatively high percentage of their sales were light trucks rather than passenger cars. Beginning in the 1970s, Detroit 3 ceded much of the U.S. passenger car market to more fuel-efficient and higher-quality vehicles manufactured by foreign-owned companies, and they found it more profitable to focus on light trucks instead. In 2007, the last year before the recession, the Detroit 3 held 52 percent of the North American market, including 65 percent of the market for light trucks but only 39 percent of the market for passenger cars. Between 2007 and 2009, sales of light trucks declined 43 percent in the United States, much steeper than the 28 percent decline for passenger cars. Many of the light trucks sold to businesses for making deliveries, hauling equipment, constructing new housing, and other activities during an expanding economy, were no longer needed during the recession. A sharp increase in gas prices during the first half of 2008 further depressed sales of gas-guzzling light trucks (Klier, 2009).

Contributing to the woes of the Detroit 3 were uncompetitive labor and legacy costs. The hourly wage rate in 2007 was $29.06 at Detroit 3 assembly plants, compared to $17.35 at all U.S. manufacturers, and $26 at U.S. assembly plants owned by Toyota, the foreign-owned manufacturer with the highest wage rates in the United States. The total hourly cost of production labor, including benefits as well as wages, was $61.48 at the Detroit 3 in 2007, compared to $47.50 at Toyota (McAlinden 2008: 20). Motor vehicle production is not a labor-intensive industry; total direct and indirect labor costs for the Detroit 3 are estimated to be around 8.4 percent of the average price of a new vehicle in the United States (UAW, 2014). But in a highly competitive market, marginal differences in labor costs can be significant.

Even more burdensome than the wages of current workers were the legacy costs, primarily generous health care obligations, paid to 700,000 Detroit 3 retirees and their dependents. In 2007, the Detroit 3 had reached an agreement to transfer responsibility for retirees' health care liabilities in 2010 to a trust, known as a voluntary employee benefits association (VEBA), administered by the United Auto Workers (UAW) union. GM agreed to provide the trust with $33 billion, Ford $15 billion, and Chrysler $9 billion (Cooney *et al.*, 2009, p.56). The 2007 labor contract introduced lower wage levels for new hires, generous buyouts that eliminated 100,000 Detroit 3 jobs between 2006 and 2010, monthly premiums for health care, co-pays for doctor visits and prescriptions, and other labor-saving components, but they were not enough to save the Detroit 3 during the severe recession (Bunkley, 2009; Vlasic, 2011).

"From a highly theoretical point of view, the correct decision could be to let Chrysler go," wrote Steven Rattner, the head of President Obama's Presidential Task Force on the Auto Industry, two years later. "[T]he substitution effect [of Chrysler customers switching to competitors' vehicles] would eventually reduce the net job losses substantially .... We intuited that the substitution analysis was more right than wrong" (Rattner, 2010: 3–4). Ultimately, though, Chrysler, as well as GM, was rescued in large measure because competing carmakers were linked to them through a common supply base. Officials of Ford, Honda, and Toyota all communicated to government officials that factories manufacturing some of their essential parts would be threatened by the collapse of one or two of their competitors. According to research published elsewhere by the author (Klier and Rubenstein, 2008), 41 percent of factories that supplied parts to one or more of the Detroit 3 also supplied parts to Asian carmakers, and 79 percent of factories that supplied parts to one or more of the North American assembly plants of Asian-headquartered carmakers also supplied parts to the Detroit 3 (Table 9.1). Note that this sharing of customers is at the level of individual factories.

Table 9.1 Extent of sharing of suppliers to Detroit 3 and international carmakers

Supplier to North American assembly plants owned by	Also supplier to North American assembly plants owned by:		
	Detroit 3	*Asia-based*	*Europe-based*
Detroit 3 carmakers	100%	41%	15%
Asia-based carmakers	79%	100%	20%
Europe-based carmakers	88%	63%	100%

Source: Klier and Rubenstein, 2008

Having determined that Chrysler and GM needed to be rescued, the Presidential Task Force concluded that the companies needed to substantially reduce the number of production facilities. North American production capacity was reduced by 31 percent at GM and by 22 percent at Chrysler. This large a reduction in production capacity could occur only through government-managed restructuring. Ford, which was not restructured by the government, also reduced its production capacity by 8 percent. Between 2008 and 2011, 16 assembly plants were closed in North America, compared to only two in Europe.

As a condition for receiving TARP loans in December 2008, the Bush Administration required Chrysler and GM to submit restructuring plans by February 17, 2009, in order to qualify for further federal assistance. President Obama announced on March 30, 2009 that the Presidential Task Force had found the initial restructuring plans inadequate. The President provided Chrysler with a final 30 days of working capital, and GM with 60 days, during which time they had to submit viable restructuring plans. Otherwise, the President warned, the companies faced bankruptcy (Webel and Canis, 2011: 6; White House, 2009, quoted in Congressional Oversight Panel, 2011: 13 and 97).

When Chrysler failed to reach restructuring agreements with all of its stake-holders within the 30-day limit, the Presidential Task Force, as a major creditor, took the company into bankruptcy court on April 30, 2009. The Presidential Task Force, which included legal experts on the U.S. Bankruptcy Code, filed under an extremely obscure section of the code known as Chapter 11 Section 363(b) that had not previously been used for manufacturers. Section 363(b) permitted the division of assets between a newly formed company and the old bankrupt company. The new company, now known as Fiat Chrysler Automobiles (FCA), took over only the assets that would permit it to operate as a viable entity, while the toxic assets remained in the old bankrupt firm for eventual liquidation or write-off (Rattner, 2010: 3, 60). Thanks to the use of Section 363(b), the "new" Chrysler exited bankruptcy protection in only 31 days, on May 31, 2009, an extremely short turnaround time for a bankrupt company, especially one of such size and complexity.

GM followed Chrysler into bankruptcy protection on June 1, 2009. The Presidential Task Force again filed under Section 363(b). Again the attractive assets were placed in a new viable company while the toxic assets remained in the old company. Again the path through bankruptcy was a remarkably short 39 days.

Policy response in Europe

On November 24, 2008, during the two-week interval between the two sets of appeals by the Detroit 3 CEOs to the U.S. Congress, French President Nicolas Sarkozy and German Chancellor Angela Merkel met in Paris to discuss the auto industry. "What is sure is that we won't let the auto industry down," promised Sarkozy at a joint press conference. "The will to help European industry, and notably the automobile sector, is total," he continued. Merkel warned that it would be "disastrous" if Europe did not respond to the deteriorating conditions in

Europe's motor vehicle industry, though she cautioned that she favored "measures that do not carry a big price tag." Though the two leaders agreed that "coordinated action was key to addressing the economic downturn," the French leader preferred a Europe-wide plan, whereas the German leader favored individual national initiatives (*The New York Times*, 2008).

Ultimately, Germany's position prevailed, as no coordinated European Union program was developed to assist the motor vehicle industry. Individual European governments treated what they regarded as a short-term cyclical downturn primarily by trying to tweak consumer demand.

Economic recovery plan

The European Commission produced several reports during the depth of the recession that addressed the motor vehicle industry either indirectly or directly (European Commission, 2008, 2009a, 2009b: 3). "A European Economic Recovery Plan," issued on November 26, 2008, two days after Merkel and Sarkozy met in Paris, listed three strategic aims for the region's overall economy, not specifically for the motor vehicle industry:

- to stimulate demand and boost consumer confidence through easing the adverse effects of the credit squeeze;
- to lessen the cost of the economic downturn on workers through improved training;
- to assist industry in speeding up the shift towards a low-carbon and resource-efficient economy required by climate change.

The Recovery Plan identified the motor vehicle industry as playing a central role in Europe's transition to a green economy. The Plan called for a €5 billion "European green cars initiative," in which the Europe Investment Bank would provide low-cost loans to motor vehicle producers and parts suppliers to finance improved safety and higher environmental standards, and governments would finance efforts to scrap old cars (European Commission, 2008).

Temporary Framework for state aid

The European Commission Communication, "Temporary Framework for state aid measures to support access to finance in the current financial and economic crisis," issued on January 22, 2009, two days after President Obama took office, authorized member states to use grants, loans, and subsidies to promote economic recovery, but the Commission warned,

> [i]n the current financial situation, Member States could be tempted to go it alone and, in particular, to wage a subsidy race to support their companies. Past experience shows that individual action of this kind cannot be effective and could seriously damage the internal market.

Europe's ailing motor vehicle industry was not mentioned specifically in the Temporary Framework, and the aid

> only applies to companies whose difficulties do not pre-date the crisis. For companies whose difficulties are mainly due to structural problems rather than the current crisis, any State aid must be supported by a restructuring plan that ensures long term viability, in line with the objectives of promoting the competitiveness of this industry.
>
> > (European Commission, 2009b)

With the authorization of the Temporary Framework, European Union members in early 2009 put in place scrappage programs, by which consumers would receive incentives for trading in older vehicles for newer ones. The details of the scrappage programs varied among European countries. Here are examples from the largest vehicle markets in Europe:

- France's *prime à la casse* (scrappage premium), introduced January 19, 2009, offered subsidies for buying a new vehicle ranging from €1,000 for a vehicle that emitted less than 160 g/km of CO_2 to €5,000 for one that emitted less than 60 g/km, while scrapping a vehicle at least 10 years old.
- Germany's *Umweltprämie* (environmental premium, popularly known as *Abwrackprämie* or scrappage premium), introduced January 13, 2009, offered a €2,500 subsidy for buying a new vehicle while scrapping a vehicle at least 9 years old.
- Italy's *incentivi alla rottamazione* (scrappage incentives) offered a €1,500 subsidy for scrapping any vehicle and buying a new vehicle that emitted less than 140 g/km of CO_2 (130 g/km for a vehicle running on diesel fuel).
- The United Kingdom's scrappage incentive scheme offered a £2,000 subsidy, funded equally by the government and the motor vehicle industry, for buying a new vehicle while scrapping a vehicle at least 10 years old.

Similar programs were enacted in other European countries. The result was an immediate increase in new vehicle registrations. In Germany, for example, new-vehicle sales were 20 percent higher in 2009 than in 2008. The increase generated false optimism. The ACEA, for example, wrote in its 2009 Economic Report that "[f]ollowing a marked decline in the second half of 2008 and first half of 2009, EU new car registrations picked up in the second half of last year [2009], reflecting the cautious upturn in the overall European economy" (European Automobile Manufacturers' Association, 2010).

Communication responding to the automotive crisis

The European Commission issued a communication specifically related to the motor vehicle industry, "Responding to the crisis in the European automotive industry," on February 25, 2009, nine days after President Obama appointed the

Presidential Task Force on the Auto Industry. "This communication sets out how the EU can bring policy levers into play to support the automotive industry. It represents a European framework for action" (European Commission, 2009a).

In reality, the communication did not offer a framework for action. Much of the communication was devoted to a reiteration of the European perspective on the issues, primarily the sharp drop in demand for vehicles brought on by lack of access to credit financing. The principal framework for action was a reiteration of early communications' emphasis on investment in green technology, as well as aid for retraining unemployed workers. Long-term structural problems of high fixed costs and overcapacity were mentioned as a contributor to the auto industry's problems, but the communication stated that

> [p]rimary responsibility for dealing with the crisis lies with industry, individual companies and their managements. Industry itself is called upon to address the structural problems of production efficiency and capacity utilisation in a way that improves its long-term competitiveness and sustainability.
> (European Commission, 2009a)

Comparing North American and European auto industry conditions 2009–2014

During the first six years following the 2009 government initiatives, motor vehicle sales and production trends varied between Europe and North America. Sales and production figures improved sharply in North America, whereas they continued to deteriorate in Europe. As a result, the North American motor vehicle industry became profitable since the government initiatives in 2009, whereas the European industry was not.

Sales and market share

North America had stronger sales and less variation among individual countries than has Europe. North America has also had less of a shift in market share.

In North America, after the sharp decline from 19.3 million in 2007 to 12.9 million in 2009, motor vehicle sales steadily recovered to 19.9 million in 2014, higher than the pre-recession year. Sales of 20.7 million in 2015 actually set an all-time record for the region. Sales during the recession declined in Europe less sharply than in North America, from 18.9 million in 2007 to 16.2 million in 2009. In contrast with North America, sales continued to decline in Europe after 2009, to 15.7 million in 2010 and 2011, 14.4 million in 2012, 14.1 million in 2013, and 13.0 million in 2014, or 69 percent of the pre-recession year, which was also the region's all-time peak year (International Organization of Motor Vehicle Manufacturers, 2015).

Since the severe recession, individual countries have displayed only slight differences in annual sales patterns within North America but considerable differences within Europe. Within North America, the United States hit an all-time sales high in 2005, declined slightly in the next two years, declined sharply during the recession, and recovered in 2015 to the highest level ever. Following a similar pattern, sales hit an all-time high in Mexico in 2006 and recovered in 2015 to an all-time high. In Canada, the drop during the recession was less severe, and historic high sales levels were achieved annually beginning in 2012.

Within Europe, national variations in sales were more severe, especially among the largest countries. In Germany, sales increased during the recession to an historic high of 4.0 million in 2009 thanks to the scrappage program, but sales declined rapidly thereafter to 3.0 million in 2014 and 3.2 million in 2015. In France, the scrappage program kept sales relatively constant at around 2.7 million during and immediately after the recession, but sales dropped rapidly to 1.8 million in 2014 and 1.9 million in 2015. In Italy, sales hit an historic peak of 2.8 million in 2007 and declined to 1.4 million in 2014 and 1.6 million in 2015. The United Kingdom, on the other hand, followed a pattern more similar to North America than to the rest of Europe, with a rapid decline during the recession, from 2.8 million in 2007 to 2.2 million in 2009, followed by an increase to 2.6 million in 2015 (*Automotive News*, various years).

The strong recovery in North America altered the long-term decline in market share of the Detroit 3 carmakers. The Detroit 3 share had fallen from 73 percent of the market in 1990 to 65 percent in 2000 and 45 percent in 2009. Between 2009 and 2015, the Detroit 3's share of the North American market remained steady at 45 percent. GM's share fell from 20 percent in 2009 to 18 percent in 2015, and Ford's declined from 16 percent to 15 percent, but Chrysler's share increased from 9 percent to 13 percent. Between 2009 and 2015, GM's North American light vehicle sales increased from 2.5 million to 3.6 million, Ford's from 2.0 million to 3.1 million, and Chrysler's from 1.2 million to 2.6 million. North America sales of Asian and European brands increased from 6.9 million in 2009 to 11.4 million in 2015 (*Automotive News*, various years).

Meanwhile, in Europe, market share trends that predated the severe recession were not altered. The three leading French- and Italian-owned carmakers—Fiat, PSA Peugeot Citroën, and Renault—continued declines that had started before the recession. The three together experienced a 16 percent decline in sales between 2007 and 2015, from 4.4 million vehicles to 3.7 million, and their combined market share in Europe declined from 31 percent in 2009 to 26 percent in 2015. In contrast, the three leading German-owned carmakers—Volkswagen, BMW, and DaimlerBenz—fared better. Their combined Europe sales increased from 4.5 million in 2009 to 5.3 million in 2015, and their share of Europe's market increased from 31 percent in 2009 to 37 percent in 2015. The carmakers with headquarters outside Europe had a mixed experience. The two U.S.-head-quartered companies—Ford and GM—declined from 19 percent of Europe's market in 2009 to 14 percent in 2015, although nearly all of Ford's decline was accounted for by selling off Jaguar, Land Rover, and Volvo during the period. Asian-headquartered companies were the main beneficiaries of increased market share in North America, but their share of Europe's sales increased only slightly, from 18 percent in 2009 to 19 percent in 2015 (*Automotive News*, various years).

Production and capacity utilization

Trends in the production of motor vehicles in North America and Europe are similar to those for sales figures, which is not surprising in view of the fact that most vehicles sold in a region are produced in the same region. As a result of the government-managed restructuring of the North American motor vehicle industry, production capacity is in line with demand for vehicles, whereas in Europe capacity remains considerably higher than demand.

Production hit its nadir in North America in 2009 at 8.8 million vehicles, down from 15.5 million two years earlier. However, production increased in the region in each subsequent year, to 12.2 million in 2010, 13.5 million in 2011, 15.8 million in 2012, 16.5 million in 2013, 17.1 million in 2014, and 17.7 million in 2015. In other words, production has recovered in North America to a record level. Meanwhile in Europe, after declining from 19.7 million in 2007 to 15.3 million in 2009, production increased to 17.2 million 2010, and 17.5 million in 2011. However, with the lingering effects of weak demand for vehicles belatedly taking their toll, production in Europe declined to 16.3 million in 2012 and 2013, before increasing to 17.1 million in 2014. Production in Europe fell below the level in North America for the first time since the 1960s.

Among individual countries within North America, variations in production around the severe recession were greater than variations in sales. In the United States, production declined between 2007 and 2009 by 47 percent, from 10.8 million to 5.7 million vehicles, and then increased between 2009 and 2014 by 109 percent, to 11.9 million vehicles. Compared with the United States, Mexico had a less severe decline during the recession of only 25 percent, from 2.1 million to 1.6 million vehicles, and a comparable recovery between 2009 and 2015 of 113 percent, to 3.5 million vehicles. On the other hand, Canada had a comparable decline during the recession of 42 percent, from 2.6 million to 1.5 million vehicles, but a more modest recovery of only 53 percent, to 2.3 million vehicles in 2015. Mexico's relatively robust and Canada's relatively weak figures are indicative of movement of an increasing share of the region's vehicle production into Mexico regardless of the recession. Mexico has become in recent years the world's leading exporter of vehicles, as a share of national production (Klier and Rubenstein, 2013a).

Production patterns also varied among individual European countries, similar to national variations in sales. Between 2007 and 2009, production declined by 32 percent in France, from 3.0 million to 2.0 million vehicles; by 34 percent in Italy, from 1.3 million to 0.8 million vehicles; and by 38 percent in the United Kingdom, from 1.8 million to 1.1 million. The decline was only 16 percent in Germany, from 6.2 million in 2007 to 5.2 million in 2009. Between 2009 and 2014, trends differed sharply among the four countries. Production continued to decline in France by 11 percent, to 1.8 million vehicles in 2014, and in Italy by 17 percent, to 0.7 million in 2014. On the other hand, production increased by 13 percent in Germany, to 5.9 million, and by 47 percent in the UK, to 1.6 million (International Organization of Motor Vehicle Manufacturers, 2015).

Although production has been more sluggish than in North America, Europe has not had a significant capacity reduction as has occurred in North America. In North America, five assembly plants were opened and 16 were closed between 2007 and 2012. During that five-year period, five plants were also opened in Europe but only two were closed. The number of North American assembly plants declined from 79 in 2003 to 73 in 2007, 70 in 2009, and 62 in 2012. By comparison, only six assembly plants in North America were shut between 1979 and 1983, around the time of the recession of the early 1980s (Klier and Rubenstein, 2013b).

The number of assembly plants operated by the Detroit 3 declined from 60 in 2003 to 52 in 2007, 46 in 2009, and 36 in 2012, as Chrysler closed three plants, Ford four, and GM nine. The three Chrysler closures were in 2008 in Fenton, Missouri, and in 2009 in Newark Delaware and a second plant in Fenton. The four Ford closures were in 2007 in Wixom, Michigan; in 2008 in Hapeville, Georgia, and Norfolk, Virginia; and in 2011 in St. Paul, Minnesota. The nine GM closures were in 2008 in Doraville, Georgia, Toluca, Mexico, and Dayton, Ohio; in 2009 in Wilmington, Delaware, Pontiac, Michigan, Oshawa, Ontario, Spring Hill, Tennessee, and Janesville, Wisconsin; and in 2012 in Shreveport, Louisiana. The Spring Hill assembly plant was reopened in 2012.

The five new plants were opened by Honda in Greensburg, Indiana, in 2009; by Kia (Hyundai) in West Point, Georgia, in 2009; by Toyota in Woodstock Ontario, in 2008 and in Blue Springs, Mississippi, in 2010; and by VW in Chattanooga, Tennessee, in 2011.

In Europe, despite the decline in sales, more assembly plants were opened than closed between 2007 and 2013. New plants were opened in 2007 by Kia in Zilina, Slovakia; in 2009 by Hyundai in Nosovice, Czech Republic; in 2012 by DaimlerBenz in Kecskemet, Hungary, and in 2013 by VW in Györ, Hungary. Two assembly plants were announced for closure: by GM in 2010 in Antwerp, Belgium; and by Fiat in 2011 at Termini in Sicily, Italy. Belatedly, carmakers in Europe have started to close more plants. Five were announced for closure in 2013 and 2014, including Ford in Southampton, England (2013), and Genk, Belgium (2014); GM in Bochum, Germany (2014); Peugeot in Aulnay-sous-Bois, France (2013); and Volvo in Uddevalla, Sweden (2013).

With more assembly plants closed since the severe recession, North America has a higher rate of utilization of plant capacity than does Europe. Capacity utilization is a key measure of the health of the industry. As a capital-intensive industry, motor vehicle production incurs considerable fixed costs of keeping open factories. Consequently, in order to be profitable, producers need to utilize their capital investment to near capacity.

A common measure of capacity utilization in the motor vehicle industry is two full-time shifts over a five-day week. If an assembly line runs at a rate of around one vehicle per minute, annual capacity is roughly one-quarter million. An assembly plant with two lines would produce one-half million vehicles a year with two fully employed shifts and no overtime. Assembly plants must utilize at least 80 percent of capacity to be profitable, according to CM-CIC Securities (*Automotive*

News Europe, 2013a). Other analysts place the break-even point a bit lower, at perhaps 75 percent (Ciferri, 2013).

According to the Federal Reserve Board, capacity utilization in the production of light vehicles in the United States averaged 77.6 percent between 1972 (when the Federal Reserve began to collect the data) and 2007. At the depth of the severe recession, during January 2009, capacity utilization of U.S. assembly plants reached a record-low level of 26 percent. After restructuring of Chrysler and GM, closure of 16 Detroit 3 assembly plants, and higher sales, capacity utilization rose from 48 percent in 2009 to 63 percent in 2010, 69 percent in 2011, 76 percent in 2012, 81 percent in 2013, and 90 percent in 2014 (Federal Reserve Bank of St. Louis, 2015).

Capacity utilization in Europe declined from 80 percent in 2007 to 57 percent in 2009, a less precipitous drop than in North America. However, capacity utilization has not rebounded in Europe as it has in North America. It has run between 61 and 65 percent between 2010 and 2013. In 2013, 58 percent of Europe's assembly plants operated at less than 75 percent capacity according to Alix Partners, and 53 percent operated below 70 percent according to IHS Automotive (Ciferri, 2013). The utilization rate varied widely within Europe. In 2012, it was estimated to be 54 percent of capacity in Italy and 60 percent in France, compared to 89 percent in Germany (Vlasic and Jolly, 2012).

The European Commission has "tried and failed" to solve the overcapacity problem, according to Philippe Jean, head of the automotive industry unit of the European Commission's Enterprise and Industry Directorate General. He blamed the failure on "lack of consensus" among member states and carmakers. "They think it's two plants from Peugeot in France and two from Fiat in Italy. No," Jean warned, "it is a broad problem felt across Europe" (Gibbs, 2013).

Outlook

Seven years after the severe recession, the future of the North American motor vehicle industry appeared bright. The government-managed restructuring during the severe recession reduced production capacity in North America to a level commensurate with demand in the region. With capacity utilization running at around 90 percent, increasing demand is likely to be met with new assembly plant construction. The increased capacity will be primarily if not entirely located in Mexico.

The U.S. government invested $10.9 billion to rescue Chrysler and ultimately recouped $9.6 billion, leaving a shortfall of $1.3 billion. Investment in GM totaled $49.5 billion, and the government ultimately recouped $39.8 billion, leaving a shortfall of $9.7 billion (*USA Today*, 2013). From that investment emerged two profitable companies. In 2013, GM earned $7.5 billion and Chrysler $1.8 billion on their North American operations. Employment in the U.S. motor vehicle industry, which had declined from 965,000 in 2007 to 625,000 in 2009, increased to 934,000 in 2015 (U.S. Bureau of Labor Statistics, 2015).

Meanwhile, prospects continued to appear grim for Europe's motor vehicle industry. Productive capacity remained much greater than forecast demand for vehicles in Europe. The projected rates of utilization of assembly plant pointed to substantial operating losses for motor vehicle producers in the region. Ford and GM, which report earnings by region, lost $1.6 billion and $800 million, respectively in Europe, in 2013, while earning $8.8 billion and $7.5 billion, respectively in North America.

Although European stakeholders and analysts agree on the magnitude of the gap between capacity and demand in the region's motor vehicle industry, they do not agree on either the cause of the gap or the appropriate corporate and public responses to closing the gap. Many analysts, especially those with international financial analysis firms, argue that the future health of Europe's motor vehicle industry requires the closure of a large number of assembly plants. Alix Partners estimated in 2012 that excess capacity in Europe was roughly 5.9 million vehicles, and a year later IHS Automotive placed the excess, including Russia, at roughly 7 million vehicles (*Automotive News Europe*, 2013a; Maynard, 2012). In 2013, CM-CIC Securities calculated that 18 assembly plants needed to be closed in Europe, and VW CEO Martin Winterkorn put the figure at ten (*Automotive News Europe*, 2013a; *Automotive News Europe*, 2013b). Laws impeding the closure of factories, strong unions, and in some cases sympathetic governments have kept open "too many" assembly plants, according to this argument.

Other Europe-based analysts assert that the gap between capacity and demand needs to be reduced through stimulating demand rather than closing factories. Government austerity measures, it is argued, have suppressed demand in Eastern Europe at the same time that production capacity has been increased there. In "normal" times, this analysis continues, the new Eastern European assembly plants would be producing vehicles to meet an increasing demand in those countries. But with demand suppressed by austerity, these factories are "dumping" vehicles into Western European countries where demand has been historically met by their own assembly plants. Gerpisa, the leading French-based auto industry analyst, points out that production and sales were roughly equivalent in Eastern Europe before the recession. For example, in 2003 production was around 1.3 million and sales around 0.9 million in Eastern Europe. But in 2011, production in Eastern Europe increased to 3.2 million while sales declined to 0.7 million. The production in Eastern Europe in excess of local demand was exported to Western Europe, to the detriment of assembly plants already there. According to Gerpisa,

> In the EU emerging markets do not emerge. Capacity created in [Eastern Europe] is used to relocate production from [Western Europe] ... not to serve local markets. If these markets had emerged, they would have profited [small car] producers such as Fiat, Renault, Peugeot, Opel and Ford, [who are] the losers today.
>
> (Julien and Pardi, 2013)

In the absence of a European Union or a coordinated multistate initiative to either reduce production capacity or stimulate sales, and a lack of agreement on whether the region's problem is primarily excess supply or suppressed demand, Europe's motor vehicle industry is likely to continue for the foreseeable future with production capacity substantially greater than sales.

Sergio Marchionne, CEO of Fiat and Chrysler, predicted in 2008 that only six major carmakers would survive the severe recession. By 2010, he told *The New York Times*, "we're going to end up with one American [carmaker], one German, of size; one European-Japanese, probably with a significant extension in the U.S.; one in Japan; one in China; and one other potential European player" (Jolly, 2008). Marchionne, recognized throughout the motor vehicle industry as a brilliant CEO, got it wrong. No major European or North American carmaker folded during or immediately after the severe recession. However, with the government-managed restructuring of the North American motor vehicle industry, the region appeared better positioned than Europe to retain its major players over the long run.

Notes

1   Unless otherwise noted, in this chapter Europe refers to EU27+ EFTA, and North America refers to Canada, Mexico, and the United States.
2   Annual statistics on motor vehicle production and sales worldwide and for individual countries are found on the web site of the International Organization of Motor Vehicle Manufacturers (www.OICA.net).
3   Auto alley is defined as the area extending south from Michigan to the Gulf of Mexico. The northern portion includes Illinois, Indiana, Michigan, and Ohio. The southern portion includes Alabama, Georgia, Kentucky, Mississippi, South Carolina, and Tennessee.
4   Canis and Webel compiled Table 2 from U.S. Treasury, Daily TARP Update, August 17, 2011 and Troubled Asset Relief Program: Monthly 105(a) Report, various dates.

References

Automotive News (annual) 'Data Center', available online: www.autonews.com.

Automotive News Europe (2008) 'Opportunities abound despite very tough market', November 10, available online: www.autonews.com/article/20081110/ANE03/811099910/0/SEARCH#axzz2kef7fong.

Automotive News Europe (2013a) 'PSA plant closing leaves Europe with 18 factories too many', October 25, available online: www.autonews.com/article/20131025/ANE/310249892/psa-plant-closing-leaves-europe-with-18-factories-too-many#ixzz2il0miXtk.

Automotive News Europe (2013b) 'VW's Winterkorn says Europe should close 10 factories', September 11, available online: www.autonews.com/article/20130911/ANE/130919970/vws-winterkorn-says-europe-should-close-10-factories#ixzz2kYZRGgwL.

Bunkley, N. (2009) 'Two in three at G.M. pass on latest buyout bid', *The New York Times*, March 26, available online: www.nytimes.com/2009/03/27/business/27auto.html.

Canis, B. and B. Webel (2011) *The role of TARP assistance in the restructuring of General Motors*, Congressional Research Service report R41978, Washington, DC: CRS Report for Congress, available online: www.fas.org/sgp/crs/misc/R41978.pdf.

Ciferri, L. (2013) 'Europe ready to re-boot', *Automotive News Europe*, September 2, available online: www.autonews.com/article/20130902/ANE/130839982/europe-ready- to-re-boot#ixzz2kSuxEHJW.

Congressional Oversight Panel (2011) *January oversight report: An update on TARP support for the domestic automotive industry*, report No. 011311, Washington, DC: U.S. Government Printing Office, available online: http://cybercemetery.unt.edu/archive/cop/20110402010325/http://cop.senate.gov/documents/cop-011311-report.pdf.

Cooney, S., J.M. Bickley, H. Chaikind, C.A. Pettit, P. Purcell, C. Rapaport, and G. Shorter (2009) *U.S. motor vehicle industry: Federal financial assistance and restructuring*, Congressional Research Service report R40003, Washington, DC: CRS Report for Congress, available online: www.fas.org/sgp/crs/misc/R40003.pdf.

European Automobile Manufacturers' Association (2010) 'Car production in 2009 at lowest level since 1996', available on-line: www.acea.be/news/news_detail/car_production_in_2009_at_lowest_level_since_1996.

European Commission (2008) 'A European economic recovery plan', communication from the European Commission COM(2008) 800, available online: http://eur-lex.europa.eu/LexUriServ/LexUriServ.do?uri=CELEX:52008DC0800:EN:NOT.

European Commission (2009a) 'Responding to the crisis in the European automotive industry', communication from the European Commission 2009/0104/final, available online: http://eur-lex.europa.eu/LexUriServ/LexUriServ.do?uri=CELEX:52009DC 0104:EN:NOT.

European Commission (2009b) 'Temporary Community Framework for state aid measures to support access to finance in the current financial and economic crisis', communication from the European Commission 2009/C 16/01, available online: http://eur-lex.europa.eu/LexUriServ/LexUriServ.do?uri=OJ:C:2009:016:0001: 0009:EN:PDF.

Federal Reserve Bank of St. Louis (2015) 'Economic research: Industrial production: Durable manufacturing: Motor vehicle and parts', available online: https://research.stlouisfed.org/fred2/series/IPG3361T3S.

Gibbs, N. (2013) 'Europe plant capacity crisis to extend to 2016', *Automotive News Europe*, June 21, available online: www.autonews.com/article/20130621/ANE/130629997/europe-plant-capacity-crisis-to-extend-to-2016#ixzz2kSuGFh1I.

Greenberg, J. (2013) 'James Carville says Barack Obama "saved the auto industry,"' *Tampa Bay Times*, November 12, available online: www.politifact.com/punditfact/article/2013/nov/12/james-carville-says-barack-obama-saved-auto-indust.

International Organization of Motor Vehicle Manufacturers (2015) 'Sales statistics' and 'Production statistics', available online: www.oica.net.

Jolly, D. (2008) 'Carmakers in Asia and Europe also look for help', *The New York Times*, December 12, available online: www.nytimes.com/2008/12/13/business/worldbusiness/13bailout.html.

Julien, B. and T. Pardi (2013) 'Why some stakeholders try to convince others that there is an American model for restructuring? The political economy of the European restructuring process', PowerPoint of paper presented at the 21st Gerpisa International Colloquium, June 13, available online: http://gerpisa.org/en/node/2181.

Klier, T. (2009) 'From tail fins to hybrids: How Detroit lost its dominance of the U.S. auto market', *Economic Perspectives*, 22:2, 2–17.

Klier, T. and J. Rubenstein (2008) *Who really made your car? Restructuring and geographic change in the auto industry*, Kalamazoo, MI: The Upjohn Institute.

Klier, T. and J. Rubenstein (2012) 'Detroit back from the brink? Auto industry crisis and restructuring, 2008–11', *Economic Perspectives*, 2012:2, 35–54.

Klier, T. and J. Rubenstein (2013a) 'The growing importance of Mexico in North America's auto production', *Chicago Fed Letter*, 310.

Klier, T. and J. Rubenstein (2013b) 'Restructuring of the U.S. auto industry in the 2008–2009 recession', *Economic Development Quarterly*, 27:2, 144–159.

McAlinden, S. (2008) 'Well, who won? The future of automotive human resources in the United States', PowerPoint presentation at the FACS Conference, Detroit, January 12.

Maynard, M. (2012) 'Why the European car market is headed for a meltdown', *Forbes*, June 28, available online: www.forbes.com/sites/michelinemaynard/2012/06/28/why-the-european-car-market-is-headed-for-a-meltdown.

The New York Times (2008) 'France and Germany pledge auto aid', November 24, available online: www.nytimes.com/2008/11/24/business/worldbusiness/24iht-24euecon.18101855.html?_r=0.

Nunnari, S. (2011) 'The political economy of the U.S. auto industry crisis', available online: www.columbia.edu/~sn2562/nunnari_autobailout.pdf.

Rattner, S. (2010) *Overhaul—An insider's account of the Obama Administration's emergency rescue of the auto industry*, New York: Houghton Mifflin Harcourt.

Spiegel On-line International (2008) 'Emergency relief: EU may be eying auto industry subsidies', October 28, available online: www.spiegel.de/international/business/emergency-relief-eu-may-be-eying-auto-industry-subsidies-a-587026.html.

UAW (2014) 'Wages and labor costs', available online: www.uaw.org/page/wages-and-labor-costs.

U.S. Bureau of Labor Statistics (2015) 'Automotive industry: Employment, earnings, and hours', available online: http://www.bls.gov/iag/tgs/iagauto.htm.

USA Today (2013) 'Feds sell $1.2 billion of GM stock, stake dwindles', November 12, available online: www.usatoday.com/story/money/cars/2013/11/12/gm-stock-sale/3511519.

Vlasic, B. (2011) *Once upon a car: The fall and resurrection of America's big three automakers—GM, Ford, and Chrysler*, New York: HarperCollins.

Vlasic, B. and D. Jolly (2012) 'Ford closing 3 plants in European downsizing', *The New York Times*, October 25, available online: www.nytimes.com/2012/10/26/business/global/ford-to-close-plants-in-britain-and-belgium-to-stem-losses.html?_r=0.

Webel, B. and B. Canis (2011) *TARP assistance for Chrysler: Restructuring and repayment issues*, Congressional Research Service report R41940, Washington, D.C.: CRS Report for Congress, available online: www.fas.org/sgp/crs/misc/R41940.pdf.

White House (2009) 'Obama administration new path to viability for GM & Chrysler', available online: www.whitehouse.gov/assets/documents/Fact_Sheet_GM_Chrysler.pdf.

10 Addressing 'strategic failure'

Widening the public interest in the UK financial and energy sectors

J. Robert Branston, Keith G. Cowling,
Philip R. Tomlinson and James R. Wilson

1 Introduction

The financial crisis of 2008 and the ensuing Great Recession (2008–2013) have led to a questioning of the dominant neo-liberal paradigm and revealed growing dissatisfaction with corporate governance (and the entrenchment of economic power) and growing inequalities within society (see, for instance, Pikety, 2014; Stiglitz, 2010, 2013). In particular, there has been widespread public discontent, not least in the UK, with the behaviour of corporate firms (and their executive directors) in strategically important monopolistic sectors, as the economy has struggled through recession.[1] In the UK, the financial sector – largely as a result of its perceived role in precipitating the crisis, its being the recipient of unprecedented state 'bailouts' and the excessive bonuses paid to corporate executives – and the energy sector – with its continual above-inflation price hikes (at a time when real wages have been falling) – have attracted specific criticism. For instance, the then Leader of the Official Opposition, Ed Miliband, had previously publicly identified both of these sectors and pledged new regulations and reform as Labour's manifesto commitments for the recent 2015 General Election (see Labour, 2015; *The Guardian*, 29/11/2013). Although Labour was unsuccessful, it is unlikely that this pledge will be entirely overlooked by his successor.

We share the concerns raised by Miliband and others, which are essentially about the concentration of monopoly capital (see Cowling and Tomlinson, 2005). However, we fear that a mere tinkering with regulation and efforts to introduce wider competition into these markets (finance and energy) is unlikely to deliver long-term outcomes that are in the 'public interest'. Rather, it is important to tackle the underlying governance failures within these (and indeed other) strategic sectors. This requires a different analysis of markets that accounts for a wider set of public interests (see Branston *et al.*, 2006). In particular, we suggest that, in contrast to traditional market-based (including 'market failure') approaches, industrial sectors should be analysed from a 'strategic choice' perspective, which not only accounts for the concentration of economic power, but also considers the implications of strategic decision-making within corporations for the wider 'public interest' (Bailey *et al.*, 2006).[2] A misalignment of the strategic interests of corporate elites and the wider public interest raises the spectre of 'strategic failure' (Cowling and Sugden, 1998). A challenge, then, is the search for (diffuse) governance structures to alleviate and negate such 'strategic failures'. This chapter adopts the strategic choice approach and does so by considering the two case studies currently attracting public attention (the financial and energy sectors), although the arguments can easily be extended to other important strategic sectors (such as health, education and transport). Thus we aim to offer an alternative interpretation of 'failures' in these sectors and then explore more inclusive ways for their future development.

The remainder of the chapter is structured as follows. In section 2, we set out the main theoretical issues, carefully defining the concepts of economic governance, strategic failure and the public interest. We then explore these concepts in the context of mainstream analyses of competition and regulation, and suggest that a strategic choice approach offers a wider perspective and a better opportunity to achieve actual outcomes that are in the public interest. Sections 3 and 4 apply the principles of the strategic choice approach to first the financial sector and then the UK energy market; in doing so, we first analyse the nature of 'strategic failure' in these sectors and then consider some ways forward. Finally, section 5 briefly concludes.

2 Theoretical issues

2.1 Economic governance, strategic failure and the public interest

The mainstream economic theory of the firm is conceptualised through a market-centred approach; this implies a market-orientated view of policy. The basis for this approach is often traced to Coase's (1937) seminal paper, where he makes the distinction between 'markets' and 'firms' as alternative spheres in the co-ordination of production. In this regard, the Coasian firm is defined by the extent to which production activity is conducted in-house without recourse to market exchange; indeed, Coase explores why markets are often 'superseded' (p. 390). This view often equates to analyses of firms' impact (in society) that are market orientated, with policy conclusions based upon correcting 'market failures' (Sugden, 1997).[3] So, for instance, in the aftermath of the financial crisis in 2007, there were calls for better regulation to correct for 'market failure' (see Beck *et al.*, 2010 and Section 3). Similarly, recent concerns over higher energy prices in the UK have led to calls for a combination of new regulation and greater competition to curb the abuse of market power. In both cases, it is market reforms and not reform of economic governance structures (within these sectors) which is the focal point of policy deliberation.

An alternative interpretation of Coase is to reprise one of the key facets of his seminal paper (1937), which was that firms are 'islands of economic planning'; that is, they plan (and undertake) economic activity (such as production, investment and employment). In modern parlance the notion of planning (by firms) equates to taking strategic decisions. These decisions determine broad business orientation and give rise to corporate strategy (see Zeitlin, 1974; Branston *et al.*, 2006).[4] Thus, the modern corporation can be seen as a locus of strategic decision-making (see Cowling and Sugden, 1987, 1998); firms are the 'means of co-ordinating production from one centre of strategic decision-making' (ibid., 1998: 67). Most critically, a strategy-focused theory of the firm has implications for economic governance. This is because, following Zeitlin (1974), to govern is to control and to control is to determine broad corporate objectives, which in its turn is to have the economic power to make strategic decisions despite resistance from other interested parties. This is important, since in contrast to mainstream interpretations, a focus upon the strategic choices of corporations (and the impacts of such decisions) makes governance a centrepiece for policy.

Within the strategic choice approach, questions arise as to the governance of modern corporations: in particular, who governs and whose (strategic) interests are prominent (and pursued). For mainstream economists, the answer is largely irrelevant; strategic decisions are taken by the firm to maximise shareholder value. For firms, employees and consumers, the free market ensures a Pareto-efficient outcome (Friedman and Friedman, 1980). In other respects, however, there has been considerable debate on differences in corporate governance structures and the role of shareholders and managers, particularly between companies and across countries, in particular between Anglo-US, Continental European and Japanese corporations (for a brief review, see Branston *et al.*, 2006).[5] However, a closer inspection suggests purported differences in corporate governance structures (and the behaviour of shareholders and managers) are relatively superficial.[6] Moreover, a consensus has emerged in recognising corporations as hierarchical entities, with *de facto* control of corporations (across the globe) resting with an elite subset of managers and executives (Branston *et al.*, 2006).

The implications of the concentration of strategic decision-making (within corporate hierarchies) are that interests vary across those concerned about, and affected by, a corporation's activities (Zeitlin, 1974; Branston *et al.* 2006). Since modern corporations are typically governed by an elite subset of those with an interest in their activities, who will be driven to pursue their own (private) strategic interests, there is a danger that the wider public interests of other stakeholders (affected by such corporate decisions) will be ignored. Cowling and Sugden (1998, 1999) refer to this as a 'strategic failure', which generates socially inefficient outcomes: 'a failure to determine the strategic direction of production ... in the broader (public) interest – in the interest of the community at large' (ibid. 1999: 361).[7] This focus upon the public interest builds upon the work of Dewey (1988) and Long (1990). For Dewey (1988), an act is private when its consequences are essentially confined to the persons directly engaged in it, but if the consequences extend beyond those directly engaged then 'the act acquires a public capacity' (p. 244). He sees (p. 257) the public as those who are 'indirectly and seriously affected for good or for evil' by an act. Drawing on this, Long views the public interest as an evolving consensus among a set of people – a public – regarding the actions of private parties; it is the standard agreed upon by that public and against which those actions can be reasonably assessed. According to him, by arriving at some consensus, a moving one, we agree on what is important both for policy and research and the latter becomes a more purposive, disciplined, cooperative endeavour as opposed to a matter of fad, fashion and funding. For public administration and political science, the appropriate standard of evaluation would appear to be the public interest.

(pp. 170–171)

We would add that it is also an appropriate standard for much of economics and economic policy. Long argues that the 'consequences of private parties'' actions create a public as that public discovers its shared concern with their effects and the need for their control; the public's shared concern with consequences is thus a 'public interest' (p. 171). Similarly, we see the public interest in a corporation's activities in general, and in its strategies in particular, as the agreed upon, evolving concerns among all of those indirectly and significantly affected by those activities and strategies (wherever they live, whatever their nationality). Thus, to avoid strategic failure, an obvious possibility is to search for ways of appropriately involving the wider publics affected by strategic decisions in the strategic decision-making process. This implies moving towards promoting more democratic governance structures, particularly within corporations, but also in socio-economic development (or territorial competitiveness) processes more widely (Cowling and Sugden, 1999; Sugden and Wilson, 2002; Wilson, 2008).[8]

2.2 Competition and regulation issues

Unfortunately, while economic governance has long been a salient issue within policy debates, a governance-centric approach to industrial economic policy has been missing from previous and contemporary practice (Cowling and Sugden, 1990, 1992; Coates, 1996). Rather, policy implementation has often focused upon reacting to and correcting 'market failures' (Pitelis, 1994), particularly in relation to competition policy (Wren, 2001), with concerns often raised in relation to the failures of government to attain outcomes preferable to those in the market (see also Coates, 1996). Such market-centred approaches give little or no direct importance to the concentration of strategic decision-making; they emphasise the benefits of markets and most especially market competition. To the extent that governance is afforded attention, the concern (of policy) is with private versus public ownership, without appreciating the crucial strategic decision-making issues that sit beneath the private/public superficialities. However, it is clear from Cowling and Sugden (1993) that ensuring the democratisation of strategic choice is not an issue that can be reduced to one of ownership.[9]

Nevertheless, there are senses in which a market-centred approach may be argued to serve the public interest, even when corporations are governed by sectional interests that are left 'free' from government 'interference'. Such arguments can be seen in extreme form in an Arrow–Debreu type general equilibrium model (see, e.g. Debreu, 1959; Arrow and Hahn, 1971). In this case, the presence of ubiquitous perfect competition ensures Pareto-optimal outcomes in situations where producers and indeed all actors pursue their own interests. Less extreme would be the situation where sectional interests govern a corporation, but where those with other interests can make takeover bids in a competitive capital market or can establish new, competing corporations. The 'strategic failure' analysis, in contrast, is founded on the significance of imperfect competition, a position justified as realistic in the context of a global economy dominated by transnational corporations and the concentration of economic power being widespread across industrial sectors (Cowling and Sugden, 1994, 1998; Cowling and Tomlinson, 2005).

Where mainstream economics recognises the presence of imperfect competition, the argument is often put forward that the public interest can be served through appropriate regulation to correct for 'market failures', which would (hopefully) yield outcomes of general benefit (see Hantke-Domas (2003) for a critical review of the 'public interest theory of regulation'). Notwithstanding the real concerns with 'regulatory capture' and the practical difficulties in regulating monopolistic firms, regulation is essentially 'an arms-length response to failures in arms-length relationships' (Branston et al., 2006; p. 203); a rigid framework is employed to enforce behaviour deduced to be in the public interest. As a result, regulation often struggles to keep pace with the evolution of the sector (as has been suggested of recent developments in the financial and energy sectors; see below). More importantly, the 'public' itself is usually divorced from the process, and there is every chance that alternatives to either the regulated outcome or the regulation mechanism might be desired. Gauging public interest through conjecture and attempting to achieve it through rigid regulation is a poor substitute for direct involvement of stakeholders in identifying and implementing optimal outcomes as an integral part of the decision-making process. A simple analogy clarifies the point. Consider the actions of parents looking after a crying baby. Since the baby can't directly communicate, the parents have to guess at the cause of the baby's crying, and it often takes time to identify the needs/desires of the child or to pacify it when these can't be identified.[10] This is akin to a regulated solution in that the benign 'regulatory authority' (the parents) are doing what they think is in the general interest of the baby (the 'public'). In contrast, when the baby is able to communicate, the parents can find out exactly what the baby is unhappy about and address it directly, often after negotiations where the opinions of baby and parents may differ. This is almost certainly a more efficient solution, in that it is quicker, because the parents don't have to go through several guesses before hitting upon a resolution, and will most likely generate a Pareto improvement in that all involved are likely to be happier with both the outcome and mechanism of getting there.

While this illustration considers mediating the interests of just one person rather than the many that form a public, the essential argument regarding communication can be extended with appropriate collective mechanisms. With respect to regulation of an industrial sector, a system based upon the articulation of the 'voice' of interested participants is likely to generate superior solutions to one which relies upon a fixed set of rules selected to achieve outcomes that are (often artificially) construed to be in the public interest.[11] Indeed, this opens an interesting bridge with a resurgent stream of literature in political science that stresses the benefits of deliberative forms of democracy (and associated mechanisms) in the determination of societal objectives (Bohman, 1998; Dryzek, 2002; Elster, 1998). It is in this essence that the strategic choice perspective advocates a governance-centric approach to policy whereby 'publics' affected by strategic decisions (by corporations and/or public agencies) are given opportunities to participate and influence the strategic decision-making process. In other words, strategic decision-making processes should be 'inclusive' and 'democratic' in arriving at decisions which are in the 'public interest' (see Branston *et al.*, 2006).

3 Strategic failure in the financial sector

3.1 Conventional analysis; market and regulatory failures

The preceding discussion is particularly relevant in the context of understanding the dynamics behind the financial crisis of 2008 (which was a major facet of the ensuing Great Recession (2008–2013) (see Branston *et al.*, 2012).[12] Conventional analyses typically focused upon 'market failures' and regulatory shortcomings that did not address (or even exacerbated) these failures. This approach apportions blame to poor internal monitoring, excessive risk-taking within financial markets and, above all, inadequate regulation to counter the range of market failures within the sector (see, e.g. Beck *et al.*, 2010; Blundell-Wignall *et al.*, 2008; Brunnermeier, 2009; HM Treasury, 2009; Krugman, 2008; Rajan, 2010; Vives, 2010). Combined with externalities in the form of co-ordination problems between actors and contagion – due to the interconnectedness of the global financial system – what began as deterioration in the 'sub-prime' lending market in the USA, rapidly transcended into a wider global financial and economic crisis (see Brunnermeier, 2009; Vives, 2010).

More generally, this complex set of issues has posed significant challenges for competition policy and regulation. Beck *et al.* (2010) document how banking regulations and competition policy evolved from the Great Depression so that, by the late 1990s, the functions of banks fundamentally changed – from being 'responsible' lenders they had greater freedom to exploit new 'financial innovations' and undertake greater risks. This meant an end to the so-called 'traditional' banking model, where funds for mortgage advances were largely sourced from retail deposits and loans advanced after a careful risk assessment of clients; such a model ensued a close monitoring of borrowers, thus guarding against adverse selection and moral hazard. To mitigate risk, loans were held as assets on balance sheets with banks holding 'appropriate' levels of capital (to meet unexpected risks) and 'loan insurance' provided through risk premia priced into the interest rate.[13] However, since the 1990s, banks have begun to finance much of their lending activities through a combination of inter-bank borrowing on global wholesale markets and securitisation (repackaging of assets to be sold on to investors as securities). In theory, the inter-bank market allows banks to easily renew their maturing borrowings to meet daily obligations. Short-term liquidity, though, varies inversely with uncertainty, and this can become a severe problem when market confidence is low. Furthermore, securitisation is a significant departure from traditional banking, where the selling of loan assets was effectively precluded due to the 'lemons problem' (Akerlof, 1970). With securitisation, the 'lemons problem' is purportedly negated by the pooling of a diverse set of loan assets through Special Purpose Vehicles (which makes them attractive to investors). These Special Purpose Vehicles were structured so that each could issue a variety of debt with different levels of agency-rated risk attached, and these risks were seen to be minimal, since insurance on default was being provided via the purchase of Credit Default Swaps. Nevertheless, potential problems arise, since the securitisation process ultimately lessens the banking sector's collective responsibility and ability to monitor loans and carefully evaluate their risk (Chick, 2008).

It is now widely accepted that these new modes of finance allowed banks to pursue aggressive strategies in lucrative mortgage markets, offering a range of low-cost products, including to customers who might be deemed to be 'risky' (Langley, 2008). A corresponding lack of prudence in lending decisions, poor monitoring of loans amidst the complexity of Special Purpose Vehicles (asymmetric information) and inappropriate risk identification were also widely acknowledged as the 'market failures' at the root of the financial crisis. Llewellyn (2009a, 2010), for instance, has argued that at the heart of the crisis in the UK was the move by bankers towards greater reliance upon wholesale funding, securitisation, the use of new financial instruments and shifting credit risk off the balance sheet. He concludes with the now universal phrase: 'banks stopped behaving like banks' (2010: 8). In addition, 'market failures' associated with co-ordination and contagion (externalities) eventually shattered confidence in the whole financial system, which spilt over into the real economy.

Not surprisingly, these 'market failures' have placed the effectiveness of (financial) regulation and the appropriate balance between stability and competition under the spotlight (see, e.g. Beck *et al.*, 2010; Blundell-Wignall *et al.*, 2008; Vives, 2010). Much of the contemporary global regulation emanates from the so-called Basel committee for international banking supervision (which among other things sets minimum capital requirements). Krugman (2008: 158), however, refers to the emergence of a 'shadow banking system' (p. 158), which bypasses much of the banking regulations and facilitates increased (bank) profitability, but without the traditional stability and safety nets.[14] Thus much blame was apportioned to the advent of unregulated financial products including auction-rate securities, collateralised debt obligations and other similar modern financial instruments (see also Brunnermeier, 2009). Further criticisms considered the mechanisms by which risk is assessed within existing regulatory frameworks, with concerns about the difficulty of statistically modelling risk due to a lack of appropriate data, and an over-reliance upon unregulated credit-rating agencies (Danielsson *et al.*, 2001; Langley, 2008).

Overall, such perspectives highlight underlying 'market failures' and, in parallel, shortcomings in financial regulation, and it is these that have informed the policy response (see, e.g. proposals for the UK as outlined in HM Treasury, 2009; Independent Commission on Banking, 2011; and the establishment of a new Basel III Accord in 2010; see also Bernanke, 2012). Such policy options tend to focus upon tightening existing financial regulations and improving monitoring to ensure more optimal handling of emerging problems.

3.2 A strategic choice analysis

While these (mainstream) analyses have provided some salient insights into the failings of the banking system, there is a danger that obsessions with 'market failure' can mask more fundamental issues. In particular, few analyses have paid significant attention to the issue of governance in the banking system and, insofar as governance has been considered, such as in the Walker Review (2009), the focus has been relatively narrow, preferring to concentrate upon failures in corporate governance within the sector.[15] In adopting a strategic choice approach, we argue that only a reform of underlying economic governance across the whole sector will deliver a policy response in the wider public interest.

Our observations begin with an interesting feature of the crisis as it afflicted the UK. This is that the three prominent retail banks which were the most adversely affected and required major intervention (and assistance, leading to – part – public ownership) by the UK Government – Northern Rock, Bradford and Bingley (B&B), and HBOS – were all former building or mutually owned societies (or were the result of mergers including former building societies), until a change in legislation in the mid-1980s allowed their conversion into (publically listed) banks.[16,17] Demutualisation had gathered pace during the 1990s, following earlier financial deregulation and increasing competition (in the building society's traditional mortgage markets) from banks and other financial institutions. This left mutuality being widely regarded as 'outdated', as building societies struggled to compete in this new environment, being compromised by the size of their capital and by legislation restricting their access to external sources of funding (such as wholesale markets). Demutualisation offered senior building society executives, who were becoming more important in shaping strategy and often had little sympathy for the principles of mutuality (and generally favoured more aggressive commercial approaches), an opportunity for greater freedom in the market (Marshall *et al.*, 1997, 2003). Supported by members' growing expectations of demutualisation windfalls, a number of leading building societies converted to public limited companies in the late 1990s. For senior executives, there were also other attractions including legal protection from hostile takeover for five years following conversion and the possibility of higher personal financial incentives and rewards, something which was instrumental in driving the demutualisation process forward. These private interests were in some cases pursued *despite resistance from others* (see also Barnes and Ward, 1999).[18]

These new banks were – following demutualisation – at the forefront of the extraordinary growth in UK mortgage lending in the 2000s. Northern Rock, for instance, had by 2007 become one of the top five mortgage lenders in the UK, with this lending largely funded from global wholesale markets (BBC News, 2007), while Bradford and Bingley relied heavily upon wholesale funding to become a specialist in the growing 'buy-to-let' market, a market with greater exposure to default, given the high number of 'self-certified' mortgages (where borrowers are not required to provide proof of income). HBOS – a merger of the former building society Halifax and the Bank of Scotland in 2001 – was the UK's largest mortgage lender and actively pursued lending in specialist higher-risk activities, such as buy-to-let and sub-prime mortgages. HBOS was also the most reliant of all UK banks upon the wholesale markets, and its lending was estimated to be almost twice that of its deposits (Sieb and Hosking, 2008).

The crisis first hit Northern Rock in September 2007, where its weaknesses led to a 'run' on the bank, the first in the UK since 1866 (Shin, 2009). It sought emergency funding (of around £3 billion) from the Bank of England and, within a week of the crisis, its share price had fallen by 60 per cent (BBC News, 2007), which weakened its position further; the consequence was an adverse credit rating, which weakened its ability to access the wholesale markets to meet maturing obligations. By January 2008, loans from the Bank of England reached £26 billion, with further guarantees of approximately £30 billion. It was taken into public ownership in February 2008. A year later, a similar fate afflicted both Bradford and Bingley and the much larger HBOS. In each case, concerns about an over-exposure to credit risk, falling share prices and the subsequent downgrading of credit ratings, prevented these banks from accessing wholesale markets; this led to a 'failure' of their business model. In September 2008, B&B was nationalised, while its savings operations were sold to the UK subsidiary of the Spanish financial group, Banco Santander. HBOS was allowed to be taken over by rival Lloyds TSB – thus (conveniently) bypassing UK competition law – so as to secure its capital base. As part of an emergency £37 billion 'bailout' of the UK banking sector, the UK government (in October 2008) took a 40 per cent share in HBOS, to become the largest single shareholder (see Branston et al., 2012).[19]

These cases reflect important concerns regarding the governance of strategic decisions. In particular, had these organisations remained mutually owned (as building societies) they *would not* have been able to (legally) pursue such risky borrowing and lending strategies. Indeed, it was noteworthy that the UK mutual sector – with one or two exceptions – weathered the recent financial storm much better than the commercial banks (Morgan and O'Hara Jakeway, 2009: 34).[20] The mutual sector – as a whole – did not move significantly away from traditional banking and was consequently better able to withstand the credit crunch (see Llewellyn, 2009b, 2010).[21] The juxtaposition of (banking) models with very different decision-making bases suggests a need to focus analysis on how processes of decision-making are articulated within banks. This suggests an analysis and policy response that is deeper than those suggested by the 'market failure approach'; the pursuit of narrow private interests – at the expense of wider public interests – is symptomatic of a more general diagnosis of 'strategic failure' within modern banking and as such requires an appropriate response.[22]

3.3   A way forward for the UK financial sector

From a strategic choice perspective, an obvious starting point for the reform of these institutions, and in the financial sector more widely, would be to consider a return to some form of mutuality. Unfortunately, the coalition and now Conservative government's ideological market-centred stance has seen moves to 'privatise' the 'bailed-out' institutions without any fundamental (governance) reform of the sector; this began with the early sale of Northern Rock and the recent partial sale (at a loss) of its stakes in Lloyds TSB and Royal Bank of Scotland. This was a missed opportunity, since reform along mutual lines has a strong theoretical appeal in maintaining a healthy balance across the sector and also on the grounds of facilitating democratic governance. In the first instance, a critical issue is the degree of risk undertaken within the overall market. Over 25 years ago, in a warning that now seems quite prescient, Llewellyn and Holmes (1991) argued that a wider variety of financial institutions provides greater stability in the overall financial architecture vis-à-vis a concentration of similar types of institutions. They argued that, since mutuals are not prone to (short-term) shareholder pressure, they were less likely to undertake risky projects; a balance of different ownership types may thus play a role in limiting (overall) instability in the event of a banking crisis. Similar arguments in favour of a mixed governance structure have been aired in the aftermath of the financial crisis; a variety of institutions with different capital structures and portfolios (i.e. 'bio-diversity'), is now regarded as being more likely to reduce systemic risk and provide for a more stable environment (Ayadi *et al.*, 2010; Llewellyn, 2009a, 2009b, 2010; Michie, 2010, 2015; Michie and Llewellyn, 2010; The Oxford Centre for Mutual and Employee-owned Business, 2009).

Second, mutuality is intuitively attractive since it implies a degree of collective ownership (among customers) which may facilitate more inclusive governance. In theory, each member can exercise their 'voice' at annual general meetings and cast equal votes on company resolutions (irrespective of the size of their custom) in a similar way to shareholders in limited companies.[23] Mutuality thus provides an environment conducive to a wider dispersion of strategic decision-making among customers (who constitute a key group of stakeholders with an interest in the activities of the organisation) (see also Marshall *et al.*, 2003). Moreover, as members are both users and (collective) owners of the business, the absence of external shareholders allows surpluses to be distributed to members in the form of low-cost mortgages and low-risk savings accounts with preferential rates of interest (Drake and Llewellyn, 2001; Kay, 1991). Indeed, many UK building societies have traditionally been involved in paternalistic activities in their own communities, often adopting 'profit-satisficing' behaviour such as maintaining (unprofitable) branch networks and extending basic financial services to financially excluded parties (see Marshall *et al.*, 2003: 743–746). Indeed, Morgan and O'Hara Jakeway (2009) have argued that the principles of mutualism 'could transform banks into the servants of their communities rather than masters of the universe' (p. 34).

Advocating a solution based upon mutuality is not, however, to deny the problems of poor governance that have long existed within the UK building society movement. These are well documented; Barnes (1984), for instance, was concerned about the monopolistic activities of the larger societies and the cartels that persisted in the 1960s and 1970s which restricted competition at the expense of members' welfare. Moreover, concerns were also raised about low levels of participation among members in annual meetings (and other decision-making processes) and the (lack of) accountability of directors to members. These issues became more acute as membership grew and senior executives became more remote. Indeed, in the more liberalised era of the 1980s and 1990s, building societies began to adopt a more commercial approach and often operated like banks, thus distancing themselves from the original ideals of mutuality (Drake, 1997).

Mutuality thus needs to be updated to avoid the problems of the past and in doing so to articulate the wider sets of publics in the society's decision-making processes. These publics would include not only customers, but also employees and perhaps representatives from the wider community, both of which are excluded from the traditional building society governance structure. In this regard, measures would need to be put in place – perhaps in the Articles of Association – to safeguard the democratic rights of all members and facilitate wider participation. It will also be important to develop appropriate mechanisms and engagement processes which can balance the different 'public' interests in the process of strategic decision-making. A real barrier to progress is that, in many mutuals, members are numerous and individually small; thus, in reality there is little member power and it is difficult to form a coalition for change (Barnes, 1984). In this regard, lessons might be learned from other member-controlled organisations such as the BBC Trust, Network Rail, Oxfam and Welsh Water. For instance, the creation of member panels might offer a way forward. The role of such member panels, which would sit between management and members, would be to actively seek and represent the views of the whole body of members and then act in a supervisory capacity (see Branston *et al.* 2012). The exact composition of members would be open for deliberation and should be appropriate for the organisation in question. It should balance the interests of different types of members with the size and nature of the firm, and also balance the need for wide participation with limits on numbers to enable a practical operational design, while ensuring that no one set of interests is unduly dominant. In this regard the panel could be established with a specific statutory duty of consulting all members on appropriate policy and reporting back to them on key decisions taken.[24] The panel's remit would need to be carefully defined to hold the management team to account and be appropriately involved with strategic decision-making, but not in a way that requires too much specialised knowledge, which would act to prohibit involvement. Thus there may be a role for specialised members who are independent of the organisation but who have the appropriate level of knowledge to inform the work of the members' panel (see Branston, *et al.*, 2012).

Such member-centric (mutual) organisations open up democratic engagement within the financial sector and offer a viable alternative to the current concentration of strategic decision-making. Moreover, an updated form of mutuality within the sector opens up further possibilities, such as mutual institutions having a specialised focus.[25] This may be a geographical or industrial focus, directing finance to developing specific regions or industries. This not only has the advantage of facilitating better consensus-building in aligning member's interests, but also in enhancing regional development, as part of a wider industrial strategy (see Bailey *et al.*, 2015).

4   Strategic failure in the British energy sector[26]

The provision of reliable and competitively priced energy is clearly important for the functioning of any modern society; electricity and gas are critical inputs for almost all industry and are relied upon for heating, lighting, cooking and for the use of many labour-saving devices and machines now taken for granted. It is therefore not surprising that post-World War II, up until the Thatcher privatisations, these services were generally provided by state-owned companies, not only in the UK, but in most other countries too.

The UK was one of the first to act when it privatised its gas industry in Great Britain with the sale of its vertically integrated provider, British Gas, in 1986. Following the lead of previous experiences in the telecoms sector, an industry-specific independent regulator, Ofgas, was created with a remit to prevent the abuse of monopoly by the newly privatised company. The vehicle of regulation was Littlechild's (1983) RPI-X price formula, where energy prices were allowed to rise (over a given period) in line with inflation less an 'X factor' deemed to reflect efficiency savings generated by the privatised company.[27] In the period since privatisation the industry has evolved considerably. Competition in supply was slowly phased in, with large buyers theoretically able to change supplier following privatisation in 1986, followed by medium-sized industrial and commercial consumers from 1992, and then by 1998 all consumers. British Gas as privatised was ultimately broken up, so that the production activities (BG Group), supply (Centrica, using the British Gas and Scottish Gas brands) and pipelines (Transco) were all demerged into separate companies (see Simmonds and Bartle, 2004). Most fundamentally though, there has been a convergence with the electricity industry.

The electricity industry and its privatisation proved to be more complex than that witnessed in the gas sector. Prior to privatisation, separate structures existed for the market in England and Wales, and for the market in Scotland, although the two markets were physically connected. In England and Wales, 12 regional Area Boards were responsible for the local distribution and supply of electricity to consumers in their monopoly area. These Area Boards purchased electricity directly from the Central Electricity Generating Board (CEGB), a vertically integrated company which was responsible for the generation and national transmission of electricity via a national electricity grid. In 1990 the British Government restructured and then subsequently privatised the electricity sector of

England and Wales. The transmission grid of the CEGB was transferred to a new company (National Grid Company – NGC) and the 12 Area Boards were essentially unchanged and became known as Regional Electricity Companies (RECs). The NGC and the RECs were seen as natural monopolies and were therefore governed by a new industry regulator, OFFER, which again employed RPI-X price regulation to protect consumers. The generation side of the CEGB was split into three companies which competed against each other to provide the electricity to the supply companies. Privatisation of the Scottish electricity sector occurred a few years later than in England and Wales, but with less restructuring. The two state-owned vertically integrated monopolies were privatised without any reorganisation except for the separation of the nuclear-generating capacity. The natural monopoly elements of the companies were again subject to regulation.[28]

As in the gas sector, there have been many developments in the electricity sector in the period since privatisation. Waterson (2015) offers a more detailed treatment of some of the changes and issues that have emerged following the privatisation of these industries (see also Branston, 2002; Simmonds, 2002). Of particular note for our purpose is that the electricity market was progressively liberalised so that supply and distribution became separated and thus so that by 1999 all consumers could also choose which firm supplied their electricity. Competition also increased in the market for generation, with far more companies being involved, either through divestment through the companies created at privatisation or via entry by new market participants. However, much of the vertical separation created at privatisation has been abandoned, and so it is now common for supply companies to own generation assets. There has also been a large amount of takeover activity such that the majority of companies involved are now part of transnational companies with global interests.

Furthermore, in the years following both privatisations there has been convergence between the gas and electricity sectors, creating what one might characterise as being a broader market for network energy. Supply companies now offer dual-fuel deals, where they look to sell customers both gas and/or electricity, and the two companies owning the national electricity transmission network in England and Wales and the high pressure main gas grid in Britain merged in 2002. To reflect such developments, the two separate industry-specific regulators merged in 1999 to form the Office of Gas and Electricity Markets, Ofgem. Given the introduction of competition in the market for supplying gas and electricity, price controls in both sectors were subsequently removed by 2002, as the regulator felt competition was sufficient to lead to appropriate price competition (see NAO, 2008). However, in the period since privatisation, there have been consistent concerns regarding 'regulatory capture' and the inefficacy of the price regulation as applied to deliver lower prices concomitant with efficiency savings (and falling marginal costs) (see for example, Branston, 2000; Waterson, 2015).

What matters for our purposes is the nature of the market (at the time of writing). Ofgem still regulates the networks used to transport the gas and electricity around the country and then locally to the end consumer, as these are still regarded as natural monopolies. Competition is relied upon to achieve the 'public interest' in the market for supplying the end consumers, and in the generation of electricity and the production/importation of natural gas. However, competition is more apparent than real as this is an oligopolistic market dominated by the 'big six' energy suppliers, who are also heavily involved in the electricity-generation element of the market. Furthermore, consumers are not particularly vigorous in regularly changing their energy supplier(s), so the market is far from being the dynamic competitive one assumed to generate ideal outcomes for consumers. Indeed the 'big six' have recently faced criticism over the profits they are earning and for being quick to raise prices when wholesale costs rise but slow to lower prices when they fall. In response, the Opposition Labour Party promised a 20-month freeze in prices if they were to win the 2015 general election, which forced the Coalition government to cut back on environmental measures in order to limit the price increases that consumers faced (*The Guardian*, 29/11/2013; BBC, 1/12/2013). Moreover, there are many other ongoing issues in the industry. These include: the type and variety of tariffs offered to consumers (which have led to many inequalities among consumers; see Waterson, 2015); the effects of green levies and other mechanisms to support the decarbonisation of the industry; the overall security of supply endangered by falling spare generation capacity and/or lack of gas storage; the 2013 deal guaranteeing an electricity price of around twice the current rate for a new nuclear electricity-generating plant to be built by EdF; and the need for significant investment to renew and decarbonise generation capacity.[29]

Given the dominance of a small number of privately owned companies earning significant profits, it is apparent that 'strategic failure' is a prevailing feature within the network energy sector. Indeed, such a conclusion was first reached in Branston *et al.* (2006) within the context of the Mexican electricity industry. In the UK, consumers have experienced increasing prices, declining security of supply leading to a real chance of service interruptions over the next few years, changes in the fuel mix used to generate electricity and increasing reliance upon imported gas. Public reaction to these changes shows that not all agree with the strategic choices behind these developments.[30] This situation has arisen because consumers have no ability to articulate their voice in the strategic decision-making processes of the industry. Indeed, for most consumers this is even more problematic because the alternative market-orientated governance process of using 'exit' to escape from an unsatisfactory state of affairs is also missing, since consumers only have access to one set of pipes or wires connecting their accommodation to the national grid and can't reasonably be expected to opt for no access to the network energy. Furthermore, other interested groups such as employees, environmentalists, domestic fuel producers and other (actual and potential) suppliers to the industry are all excluded from the decision-making processes that are currently dominated by a subset of shareholders in the dominant companies.

There is also 'strategic failure' in a wider sense. Significant parts of the network energy market, most especially those regarded as being natural monopolies, are subject to regulation by Ofgem; the approach of this regulation is unsurprisingly based upon mainstream thinking, with a focus on trying to protect the consumer by mimicking the effects of competition where competition isn't feasible. Consultations with interested parties take place, but key strategic decisions are taken by a subset of individuals with an interest (within the confines of the regulation rules agreed by government) and by those appointed to regulate the industry by government. This is a very narrow interpretation of the 'public interest' and clearly prevents all those with an interest being part of the decision-making process and truly articulating their voice (see www.ofgem.gov.uk – last accessed 7 May 2014).

4.1   A way forward for the UK energy sector

One way of solving many of the immediate issues within the energy market would be to return the industry to a form of social ownership (e.g. renationalisation). While other European economies continue to have state-owned energy companies – most notably France – the practicalities of returning these privatised utilities back to the state sector within the UK are difficult and currently (short of a revolution!) unrealistic, given policymakers' current concerns over the size of the government's budget deficit and the significant compensation that would be required to reimburse shareholders for the assets of the energy sector. Moreover, significant investment is required in the next few years within the sector to meet the UK's future energy requirements, and while the state could theoretically finance such investment, there is little political appetite to do so. Moreover, renationalisation might be complicated in light of current EU competition regulations, particularly in relation to state aid (Department of Energy and Climate Change, 2013).

More fundamentally, renationalisation (by itself) does not address the issues with governance that are the root cause of the strategic failure highlighted above. An approach is therefore needed that sits within the wider policy constraints but does so whilst also creating appropriate governance structures that give 'voice' to the different publics with an interest so that they might articulate their views.[31] In this regard, our earlier analysis on the financial sector is insightful. The government might facilitate the creation of new supply and electricity generation companies of a significant size to change the way in which the market currently functions.[32] These new companies might be newly created, or might be expanded and reformed versions of smaller existing companies with suitably constructed governance mechanisms. Such entry would increase competition in the market, appealing to advocates of a mainstream solution centred upon competition, but crucially these new companies could be run in the wider public interest rather than in the interest of a subset of private shareholders. They might be mutually owned and structured as per the reformed building societies mentioned in section 3, or they might be something more akin to Companies Limited by Guarantee such as Network Rail and Welsh Water, where there are professional managers held to account by a small group of members who are appointed to represent the wider public. There could again be particular constituencies for different types of member, including those with more technical knowledge if appropriate. That such companies could operate alongside their commercial rivals is feasible; indeed a small niche of energy supply companies already compete against the 'big six'. Moreover, like in the financial sector, a variety of firms operating in the sector would be useful for the sector's robustness and provide consumers with a real choice (where price would be just one aspect of the choice of supplier). Such new companies would be able to respond to the 'public interest' and so might differentiate themselves in many ways, including the number and type of tariffs they offer, the localities they operate in, the type of products they offer (e.g. locally produced electricity from green or other sources), and indeed, any other way that 'publics' identify.

In order to ensure such firms can fairly compete in the market, supportive wider reforms are needed. This could include measures to make sure the entire network energy industry is once again vertically disintegrated so that supply companies do not own production assets which they favour, thereby helping to create a more competitive market for electricity generation.[33] Green levies and other governmental targets might also need to be modified to make sure that these do not disadvantage any one particular company whilst still allowing wider decarbonisation goals to be achieved.

A further complementary policy would be to reform the regulation that looks after the sector, most pertinently the natural monopoly elements. In an ideal world the governance of the national grid company, like all other firms, would also be reformed along the lines suggested above, but this would be difficult in light of the fact that they are already a large private sector company owned by shareholders. As a result, an immediate focus could easily be placed upon industry-specific regulation. The effectiveness of the current regime has been questioned, given the profitability of the companies in question and rising energy prices. Irrespective of the economic appropriateness of the current regime, what is beyond doubt is that the governance of Ofgem is deficient. At the moment the regulators are appointed by the government and are tasked with a pure economic competition mandate within some environmentally set constraints. In order to address this, the governance of Ofgem might be reformed so that it becomes more able to respond to broader public interests. This cannot be achieved simply through a fixed change in the rules of regulation (as mainstream thinking suggests), but requires that the public are directly able to feed into its decision-making so that they can exercise 'voice'. One possibility is to create a wide-ranging public interest supervisory board within Ofgem, where different interest groups could be represented and where strategic issues of price are balanced with the other, wider concerns of the public. Such a board might be characterised as being akin to the panels being recommended for reformed building societies, as mentioned earlier. Thus it would be different from the current Gas and Electricity Markets Authority, which is essentially the Ofgem equivalent of a corporate board, where appointments are made by the Secretary of State at the Department of Energy and Climate Change, and where members are appointed because they are experts in the technical areas of regulation (see www.ofgem.gov.uk – last accessed 7 May 2014). The crucial point is that such member-centric reform to the regulator would open up democratic engagement within all aspects of the energy sector and would therefore offer a viable alternative to the current concentration of strategic decision-making that is so apparent within the sector.

5   Conclusion

In this chapter, we have raised fundamental concerns with the economic governance of two strategically important sectors for the UK (and indeed any other) economy: the financial and energy sectors. We have documented how the concentration of economic power and strategic decision-making processes within these sectors has led to the spectre of 'strategic failure'. Indeed, both sectors have recently been the focus of much public discontent. Unfortunately, both current and proposed regulations are ineffective in delivering these (and indeed other) public services in the public interest.

What is required is a governance-centric approach that places inclusion and democratic choice as the focal point of public policy. Adopting such an approach, we have outlined some realistic and feasible possibilities for each of these sectors, which could aid their future development in the wider public interest. Moreover, such proposals could be adapted for other sectors in a wider move to transform the economy towards a more sophisticated form of economic democracy that is more responsive to a broader range of public interests and less likely to be captured by small elites of disproportionately powerful interests. Indeed, we suggest that policy-makers would do well to explore such possibilities if they are serious about confronting monopoly power.

Notes

1   Indeed, new evidence suggests that in the UK and the US, monopolistic practices tightened during the recession, as firms recognised their mutual adversity and tacitly raised price-cost margins to the detriment of the public (see Branston *et al.*, 2014).

2   As will become apparent in Section 2, the strategic choice approach has its roots in Cowling and Sugden's (1987, 1994, 1998, 1999) analysis of the modern corporation and their relationships with national and local economies.

3   In Williamson's (1975, 1985) adaptation of Coase's work, firms (and their internal hierarchies) are seen as economising upon the transaction costs associated with market exchange; thus these transaction cost-saving activities are regarded as 'Pareto efficient'. This point is fiercely contended by Cowling and Sugden (1998), on the grounds that organisational changes (such as shifting from production based upon market exchange to internal hierarchy) invariably incurs distributional consequences, thus disallowing Pareto efficiency.

4   In line with Zeitlin (1974), the power to make strategic decisions includes the ability to plan a corporation's relationships with other corporations, its relationships with governments and employees, and its geographical orientation.

5   See, for example, Aoki (1990), Jenkinson and Mayer (1992), Shleifer and Vishny (1997), Bechtand and Röell (1998), Scott (1999), Yafeh (2000).

6   For instance, Coffey and Tomlinson (2003a, 2003b) provide a strong rebuttal of Aoki's (1990) position on the (perceived) differences in the governance of Japanese (the non-hierarchical – J mode) vis-à-vis Western (hierarchical – H mode) corporations. Using data on the automobile sector (Aoki's example), they deconstruct Aoki's

J mode firm, to demonstrate that Japanese firms are also hierarchical entities.

7   We have previously documented several cases of 'strategic failure' across the globe (see, for instance, Cowling and Tomlinson, 2000, 2005, 2011a, 2011b; Branston *et al.*, 2012).

8   Evidence in Frey and Stutzer (2000: 918) suggests that widening direct involvement in strategic decision-making might in itself increase people's well-being. According to evidence from interview data in Switzerland, 'individuals are ceteris paribus happier, the better developed the institutions of direct democracy are in their area of residence'.

9   Ownership *per se* is not the issue; it is important only in so far as different ownership settings in different contexts may render more or less likely the participation of those with an interest ('publics') in a corporation's decision-making processes. Indeed, this point seems to have been recognised by the recently established UK Commission on Ownership (see http://ownership-comm.org – last accessed 7 May 2014 – for more information).

10 We know from personal experience!

11 The baby analogy is useful for a further point regarding regulation. Sometimes the public is 'wrong' and regulation is needed to limit what they would otherwise like to happen. It isn't always appropriate for a child to stay up late and eat chocolate. The key difference in this situation is that utilising regulation alone removes (or at the very least reduces) the possibility of the public themselves identifying superior solutions or outcomes. The public in this setting has no possibility of using Hirschman's (1970) concept of voice.

12 For an extensive overview of the events of the financial crisis, see Brunnermeier (2009). There are, however, wider interpretations of the underlying causes of the Great Recession; Cowling *et al.* (2011), for instance, provide an analysis that points to the unsustainability of 'global imbalances' between economies, which underpinned much of the ensuing distress within the financial system. They argue that these 'global imbalances' were very much the consequence of economies having predominantly monopolistic structures.

13 The performance of banks reflected the quality of their own lending decisions, which was their own responsibility (see Michie and Llewellyn, 2010).

14 Bernanke (2012) defines 'Shadow banking as comprising a diverse set of institutions and markets that, collectively, carry out traditional banking functions – but do so outside, or in ways only loosely linked to, the traditional system of regulated depository institutions. Examples of important components of the shadow banking system include securitization vehicles, asset-backed commercial paper (ABCP) conduits, money market mutual funds, markets for repurchase agreements (repos), investment banks, and mortgage companies'.

15 The Walker Review (2009) considered corporate governance of UK banks and financial enterprises, wryly noting that 'better financial regulation cannot alone satisfactorily assure performance of the major banks … these entities need to be better governed' (p. 9). The Review goes on to make a series of recommendations for improving governance structures. These proposals primarily related to the composition of the boards of directors (and the role of chairpersons) and to encouraging greater activism on the part of non-executive directors and institutional shareholders in monitoring and engaging with corporate strategy. The review made 38 recommendations in relation to reforming corporate governance within the UK banking and finance sector. Further details are available at www.hm-treasury.gove.uk/walker_review_information.htm (last accessed 7 May 2014). While many of these recommendations are sensible ways forward, it is doubtful these alone are sufficient to avert the risk strategies adopted by the modern-day banks. Indeed, it was notable that, in a climate of rising property prices and a belief in rational expectations (and efficient markets), institutions such as rating agencies, central banks, governments

and others failed to question the strategies and business operations of banks in the years leading up to the crisis (Llewellyn, 2010).

16   A 'building society' is a mutually owned bank which specialises in mortgages and saving products.

17   It is also noteworthy that, in addition to these three cases, not one of the other demutualised building societies has survived as an independent financial institution, the others being subsequently taken over by rival banking groups (Michie and Llewellyn, 2010).

18   See, for instance, example, Perks's (1991) detailed account of the measures taken by Abbey National's corporate executives to nullify opposition to that society's demutualisation process in the late 1980s.

19   To comply with EU state aid rules, the enlarged Lloyds Banking Group was required to divest a significant package of branches, brands and customer accounts (European Commission, 2009). It was announced in December 2011 that the mutually owned Co-operative Banking Group was the preferred buyer of these assets. This proved problematic, as the Co-operative Bank was unable to complete the deal – due its own problems (see footnote 22). In 2013, TSB was split from the Lloyds Banking group to meet EU competition rules (see *The Daily Telegraph*, 07/09/2013).

20   Due to legal restrictions on accessibility to wholesale funding, reliance upon internal capital and a generally lower attitude to risk, the mutual sector was by and large far more prudent in its business decisions.

21   However, the mutual sector was adversely affected by contagion effects such as falling house prices and increasing personal bankruptcies.

22   Indeed, the difficulties of a few building societies (such as the Dunfermline Building Society) and the Co-operative Bank can be attributed to a lack of accountability in their governance structures and associated behaviour outside their traditional domains in attempts to compete in certain markets (see Branston *et al.*, 2012).

23   The 'exit' option is also available by moving deposits elsewhere.

24   Seats on the panel could, for instance, be reserved for borrower members, saver members and employees, and perhaps others from the wider community.

25   Any new mutual society will continue to face competitive pressures to act commercially, although if structures, regulations and Articles of Association are designed carefully, then there is no reason why mutuals cannot co-exist with more commercial organisations, as they have done for more than 150 years previously, and pursue wider interests than profit. In addition, there may be pressures for a further demutualisation (which is allowed under current legislation). To ensure stability and safeguard against this, some form of 'asset lock' preventing members from realising the mutual's underlying assets might be put in place. Charitable assignment practices have so far proved to be a successful defence (against demutualisation) in existing building societies (Michie and Llewellyn, 2010).

26   There is distinct gas and electricity provision in Northern Ireland and so what is discussed herein refers to Great Britain and not to the UK as a whole – see Simmonds (2002), Simmonds and Bartle (2004), and the Northern Ireland Utility Regulator (www.uregni.gov.uk/about_us – last accessed 7 May 2014) for more detail.

27   In some cases, the formula became more complex to allow privatised firms the opportunity to recoup the fixed costs of new investments (which were required to improve service delivery) (see Waterson, 2015).

28   For more on electricity privatisation see Chesshire, 1996; Thomas, 1996; Newbery and Green, 1996; Branston, 2002; and Simmonds 2002.

29   For instance, it has been estimated that £110 billion is required to create a low-carbon and reliable electricity industry (HM Government, 2011).

30   Moreover, at the same time, the share prices of the privatised energy companies have risen dramatically, reflecting their high profitability. For instance, the share price of British Gas (and successor companies) has been estimated to have risen by 1,822 per cent since privatisation in 1986 (*The Guardian*, 11/10/13).

31 One might think of such constraints including, amongst other things, governmental environmental levies and associated goals on the decarbonisation of the industry, the increasingly international nature of infrastructure investment opportunities and funding sources, and wider industrial policy objectives such as using (or not) domestically produced fuel or equipment.

32 The national transmission and local distribution of both electricity and gas are both still natural monopolies, and hence entry by new firms is not possible due to the need to obtain economies of scale. Domestic production and the importation of natural gas appears to be more competitive, given the global pricing that exists for such a product, and hence the possibilities are consequentially less obvious.

33 See Waterson (2015) and also Branston (2002) as to the issues that supply companies owning generation capacity has created.

References

Akerlof, G.A. (1970) 'The market for "lemons": quality uncertainty and the market mechanism', *Quarterly Journal of Economics*, 84: 488–500.

Aoki, M. (1990) 'Toward an economic model of the Japanese firm', *Journal of Economic Literature*, 28 (1): 1–27.

Arrow, K.J. and Hahn, F.H. (1971) *General Competitive Analysis*, San Francisco, CA: Holden-Day.

Ayadi, R., Llewellyn, D. T., Schmidt, R. H., Arbak, E. and De Groen, W. P. (2010) *Investigating Diversity in the Banking Sector in Europe: Key Developments, Performance and Role of Cooperative Banks*, Brussels: Centre for European Policies Studies. Available at: www.ceps.be/book/investigating-diversity-banking-sector-europe-key-developmentsperformance-and-role-cooperative (last accessed 7 May 2014).

Bailey, D., De Propris, L., Sugden, R. and Wilson, J. R. (2006) 'Public policy for European economic competitiveness: an analytical framework and a research agenda', *International Review of Applied Economics*, 20: 555–572.

Bailey, D, Cowling, K. and Tomlinson, P.R. (2015) 'An industrial strategy for UK cities' in Bailey, D., Cowling, K. and Tomlinson, P.R. (eds) *New Perspectives on Industrial Policy for a Modern Britain*, Oxford: Oxford University Press (Edited Volume), pp. 263–286.

Barnes, P. (1984) *Building Societies: The Myth of Mutuality*, London: Pluto Press.

Barnes, P. and Ward, M. (1999) 'The consequences of deregulation: a comparison of the experience of UK building societies with those of the US savings and loan associations', *Crime, Law and Social Change*, 31: 209–244.

BBC (2013) 'Government outlines plans to cut energy bills by £50', BBC news online, 1 December. Available at: www.bbc.co.uk/news/uk-25170774 (last accessed 7 May 2014).

BBC News (2007) 'Rush on Northern Rock continues', 15 September. Available at: http://news.bbc.co.uk/1/hi/business/6996136.stm (last accessed 7 May 2014).

Becht, M. and Röell, A. (1998) 'Blockholdings in Europe: an international comparison', *European Economic Review*, 43 (4–6): 1049–1056.

Beck, T., Coyle, D., Dewatripont, M., Freixas, X. and Seabright, P. (2010) *Bailing Out the Banks: Reconciling Stability and Competition. An Analysis of State-Supported Schemes for Financial Institutions*, London: Centre for Economic Policy Research.

Bernanke, B.S. (2012) 'Reflections on the crisis and the policy response', at the *Russell Sage Foundation and The Century Foundation Conference on 'Rethinking Finance,'* New York, 13 April.

Blundell-Wignall, A., Atkinson, P. and Lee, S.H. (2008) 'The current financial crisis: causes and policy issues', *Financial Market Trends*, 96: 11–31.

Bohman, J. (1998) 'Survey article: the coming age of deliberative democracy', *The Journal of Political Philosophy*, 6: 400–425.

Branston, J.R. (2000) 'A counterfactual price analysis of British electricity privatisation', *Utilities Policy*, 9 (1): 31–46.

Branston, J.R. (2002) 'The price of independents: an analysis of the independent power sector in England and Wales', *Energy Policy*, 30 (15): 1313–1325.

Branston, J.R., Cowling, K. and Sugden, R. (2006) 'Corporate governance and the public interest', *International Review of Applied Economics*, 20 (2): 189–212.

Branston, J.R., Tomlinson, P.R. and Wilson, J.R. (2012) '"Strategic failure" in the financial sector: a policy view', *International Journal of the Economics of Business*, 19, (2): 169–189.

Branston, J.R., Cowling, K. and Tomlinson, P.R. (2014) 'Profiteering and the degree of monopoly in the Great Recession: recent evidence from the US and the UK', *Journal of Post Keynesian Economics*, forthcoming.

Branston, J.R., Sugden, R., Valdez, P. and Wilson, J.R. (2006) 'Generating participation and democracy: an illustration from electricity reform in Mexico', *International Review of Applied Economics*, 20 (1): 47–68.

Brunnermeier, M. K. (2009) 'Deciphering the liquidity and credit crunch 2007–2008', *The Journal of Economic Perspectives*, 23: 77–100.

Chesshire, J., 1996. 'UK electricity supply under public ownership, in J. Surrey (ed.) *The British Electricity Experiment. Privatization: The Record, the Issues, the Lessons*, London: Earthscan, pp.14–39.

Chick, V. (2008) 'Could the crisis at Northern Rock have been predicted?: an evolutionary approach', *Contributions to Political Economy*, 27: 115–124.

Coase, R.H. (1937) 'The nature of the firm', *Economica*, 4: 386–405.

Coates, D. (1996) 'Introduction', in D. Coates (ed.) *Industrial Policy in Britain*, London: Macmillan.

Coffey, D. and Tomlinson, P.R. (2003a) 'Co-ordination and hierarchy in the Japanese firm: the strategic decision-making approach vs Aoki', in M. Waterson (ed.) *Competition, Monopoly and Corporate Governance*, Cheltenham: Edward Elgar Publishing, pp. 3–19.

Coffey, D. and Tomlinson, P.R. (2003b) 'Globalisation, vertical relations and the J-mode firm', *Journal of Post Keynesian Economics*, 26 (1): 117–144.

Cowling, K. and Sugden, R. (1987) *Transnational Monopoly Capitalism*, Sussex: Wheatsheaf Books.

Cowling, K.G. and Sugden, R. (1990) *A New Economic Policy for Britain: Essays on the Development of Industry*, Manchester: Manchester University Press.

Cowling, K.G and Sugden, R. (1992) *Current Issues in Industrial Economic Strategy*, Manchester: Manchester University Press.

Cowling, K. and Sugden, R. (1993) 'A strategy for industrial development as a basis for regulation', in R. Sugden (ed.) *Industrial Economic Regulation. A Framework and Exploration*, London: Routledge. Reprinted in D. Parker (ed.) (2000) *Privatization and Corporate Performance*, Cheltenham: Edward Elgar.

Cowling, K., and Sugden, R. (1994) *Beyond Capitalism: Towards a New World Economic Order*, London: Pinter.

Cowling, K. and Sugden, R. (1998) 'The essence of the modern corporation: markets, strategic decision-making and the theory of the firm', *The Manchester School*, 66: 59–86.

Cowling, K. and Sugden, R. (1999) 'The wealth of localities, regions and nations: developing multinational economies', *New Political Economy*, 4: 361–378.

Cowling, K. and Tomlinson, P.R. (2000) 'The Japanese crisis: a case of strategic failure?', *The Economic Journal*, 110 (464): F358–F381.

Cowling, K. and Tomlinson, P.R. (2005) 'Globalisation and corporate power', *Contributions to Political Economy*, 24 (1): 33–54.

Cowling, K. and Tomlinson, P.R. (2011a) 'Post the 'Washington consensus: economic governance and industrial strategies for the 21st century', *Cambridge Journal of Economics*, 35 (5): 831–852.

Cowling, K. and Tomlinson, P.R. (2011b) 'The Japanese model in retrospective: industrial strategies, corporate Japan and the "hollowing out" of Japanese industry', *Policy Studies*, 32 (6): 569–583.

Cowling, K., Dunn, S. and Tomlinson, P.R. (2011) 'Global imbalances, the present crisis and modern capitalism: a structural approach', *Journal of Post Keynesian Economics*, Summer, 33 (4): 575–600.

Danielsson, J., Embrechts, P., Goodhart, C., Keating, C., Muennich, F., Renault, O. and Shin, H.S. (2001) 'An academic response to Basel II', special paper no. 130, LSE Financial Markets Group. Available at: www.bis.org/bcbs/ca/fmg.pdf (last accessed 7 May 2014).

Debreu, G. (1959) *Theory of Value: An Axiomatic Analysis of Economic Equilibrium*, New York: Wiley.

Department of Energy and Climate Change (2013). *The Energy Review: Call for Evidence on the Government's Review of the Balance of Competences between the United Kingdom and the European Union*. Available at www.gov.uk (last accessed 7 May 2014).

Dewey, J. (1988) [1927] *The Public and its Problems*, 1927: Denver, CO: Holt. Page numbers refer to the reproduction in Boydston, J.A. (ed.) (1988) *John Dewey. The Later Works Volume 2: 1925–1927*, Carbondale, IL: Southern Illinois University Press.

Drake, L. (1997) 'The economics of mutuality', LUBC-BSA Project Paper 3, London: The Building Societies Association.

Drake, L. and Llewellyn, D.T. (2001) 'The economics of mutuality: a perspective on UK Building Societies', in Johnston Birchall (ed.) *The New Mutualism in Public Policy*, London: Routledge, pp. 14–40.

Dryzek, J. (2002) *Deliberative Democracy and Beyond: Liberals, Critics, Contestations*, Oxford: Oxford University Press.

Elster, J. (1998) 'Introduction', in Jon Elster (ed.) *Deliberative Democracy*, Cambridge: Cambridge University Press, pp. 1–18.

European Commission (2009) 'State aid: Commission approves restructuring plan of Lloyds Banking Group', press release database, IP/09/1728, Brussels.

Friedman, M. and Friedman, R. (1980) *Free to Choose: A Personal Statement*, New York: Harcourt Publishers.

Frey, B. and Stutzer, A. (2000) 'Happiness, economy and institutions', *The Economic Journal*, 110: 918–938.

Hantke-Domas, M. (2003). 'The public interest theory of regulation: non-existent or misinterpretation?', *European Journal of Law and Economics*, 15: 165–194.

Hirschman, A.O. (1970) *Exit, Voice and Loyalty: Responses to Decline in Firms, Organizations and States*, Cambridge, MA: Harvard University Press.

HM Government (2011) 'Planning our electric future: a white paper for secure, affordable and low carbon electricity', London: Department for Energy and Climate Change, The Stationery Office.

HM Treasury (2009) 'Reforming financial markets', CM7667, London: The Stationary Office. Available at: www.hm-treasury.gov.uk/reforming_financial_markets.htm (last accessed 7 May 2014).

Independent Commission on Banking (2011) 'Final report: recommendations'. Available at: www.webarchive.national.gov.uk (last accessed 30 November 2015).

Jenkinson, T. and Mayer, C. (1992) 'The assessment: corporate governance and corporate control', *Oxford Review of Economic Policy*, 8 (3): 1–10.

Kay, J. (1991) 'The economics of mutuality' *Annals of Public and Co-operative Economics*, 62: 309–318.

Krugman, P. (2008) *The Return of Depression Economics and the Crisis of 2008*, London: Penguin Books.

Labour (2015) 'Britain can be better. The Labour Party manifesto 2015'. Available at: www.labour.org.uk/manifesto (last accessed 30 November 2015).

Langley, P. (2008) 'Sub-prime mortgage lending: a cultural economy', *Economy and Society*, 37: 469–494.

Littlechild, S.C. (1983) *Regulation of British Telecommunications' Profitability*, Report to the Secretary of State for Industry, London: HMSO.

Llewellyn, D.T. (2009a) 'The new banking and financial system', in David Mayes, Robert Pringle and Michael Taylor (eds) *New Frontiers in Regulation and Oversight of the Financial System*, London: Central Banking Publications, pp. 49–58.

Llewellyn, D.T. (2009b) 'Building societies and the crisis'. *Butlers Building Society Guide*, London: ICAP.

Llewellyn, D.T. (2010) 'Banking after the crisis: come back Captain Mainwaring – all is forgiven?' *Butlers Building Society Guide*, London: ICAP.

Llewellyn, D.T. and Holmes, M.J. (1991) 'In defence of mutuality: a redress to an emerging conventional wisdom', *Annals of Public and Co-operative Economics*, 62: 319–354.

Long, N.E. (1990) 'Conceptual notes on the public interest for public administration and policy analysts', *Administration and Society*, 22 (2): 170–181.

Marshall, J.N., Richardson, R., Raybould, S. and Coombes, M. (1997) 'The transformation of the British building society movement: managerial divisions and corporate reorganisation, 1986–1997', *Geoforum*, 28: 271–288.

Marshall, J. N., Willis, R. and Richardson, R. (2003) 'Demutualisation, strategic choice, and social responsibility', *Environment and Planning C: Government and Policy*, 21: 735–760.

Michie, J. (2010) 'Promoting corporate diversity in the financial services sector', research paper for the commission on ownership. Available at: http://ownershipcomm.org/research (last accessed 7 May 2014).

Michie, J. (2015) 'Financial architecture and industrial policy: alternative ownership structures and the role of mutuals', in Bailey, D., Cowling, K. and Tomlinson, P.R., *New Perspectives on Industrial Policy for a Modern Britain*, Oxford: Oxford University Press.

Michie, J. and Llewellyn, D.T. (2010) 'Converting failed financial institutions into mutual organisations', *Journal of Social Entrepreneurship*, 1: 46–170.

Morgan, K. and O'Hara Jakeway, J. (2009) 'Mutualism – an idea whose time has come (again)', in John Tomaney (ed.) *The Future of Regional Policy*, London: The Smith Institute, pp. 34–39.

National Audit Office – NAO (2008) *Protecting consumers? Removing Retail Price Controls*, report by the Comptroller and Auditor General, NC 342 Session 2007–2008.

Available at: www.nao.org.uk/wp-content/uploads/2008/03/0708342.pdf (last accessed 7 May 2014).

Newbery, D. and Green, R. (1996) 'Regulation, public ownership and privatisation of the English electricity industry', in R.J. Gilbert and E.P. Kahn (eds) *International Comparisons of Electricity Regulation*, New York: CUP, pp. 25–81.

Perks, R.W. (1991) The fight to stay mutual: Abbey members against flotation versus Abbey National Building Society, *Annals of Public and Co-operative Economics*, 62: 393–429.

Pikety, T. (2014) *Capital in the Twenty-first Century*, Harvard, CT: Harvard University Press.

Pitelis, C. (1994) 'Industrial strategy: for Britain, in Europe and the world', *Journal of Economic Studies*, 21 (5): 3–92.

Rajan, Raghuram G. (2010) *Fault Lines: How Hidden Fractures Still Threaten the World Economy*, Princeton, NJ: Princeton University Press.

Scott, K. (1999) 'Institutions of corporate governance', *Journal of Institutional and Theoretical Economics*, 155 (1): 3–21.

Shin, H.S. (2009) 'Reflections on Northern Rock: the bank run that heralded the global financial crisis', *The Journal of Economics Perspectives*, 23: 101–119.

Sieb, C. and Hosking, P. (2008) 'HBOS shares slump puts the frighteners on 15 million savers', *The Times*, 17 September.

Simmonds, G. (2002) *Regulation of the UK Electricity Industry*, CRI industry brief, Bath: University of Bath. Available at: www.bath.ac.uk/management/cri/pubpdf/Industry_Briefs/Electricity_Gillian_Simmonds.pdf (last accessed 7 May 2014).

Simmonds, G. and Bartle, I. (2004) *The UK Gas Industry*, CRI industry brief, Bath: University of Bath. Available at: www.bath.ac.uk/management/cri/pubpdf/Industry_Briefs/Gas_Brief_Ian%20Bartle_Gillian_Simmonds.pdf (last accessed 7 May 2014).

Shleifer, A. and Vishny, R.W. (1997) 'A survey of corporate governance', *Journal of Finance*, 52 (2): 737–783.

Stiglitz, J. (2010) *Freefall: The Fall of Free-markets and the Sinking of the Global Economy*, New York: Penguin Books.

Stiglitz, J. (2013) *The Price of Inequality*, New York: Penguin Books.

Sugden, R. (1997) 'Economías multinacionales y la ley del desarrollo sin equidad (Multinational economies and the law of uneven development)', *FACES Revista de la Facultad de Ciencias Económicas y Sociales*, 3 (4): 87–109.

Sugden, R, and Wilson, J.R. (2002). 'Economic development in the shadow of the consensus: a strategic decision-making, *Contributions to Political Economy*, 21: 111–134.

The Daily Telegraph (07/09/2013) 'Welcome back, TSB, as bank splits from Lloyds', London.

The Guardian (11/10/2013) 'Royal Mail shares soar 38% as Labour complains of a knockdown price', London.

The Guardian (29/11/2013) 'Energy bills: Ed Miliband promises simplification on top of price freeze', *The Guardian*, London.

The Oxford Centre for Mutual and Employee-owned Business (2009) *Converting Failed Financial Institutions into Mutual Organisations*, Kellogg College, University of Oxford. Available at: www.kellogg.ox.ac.uk/researchcentres/documents/Converting%20financial%20institutions%20into%20mutual%20organisations%20-%20final.pdf (last accessed 7 May 2014).

The Walker Review (2009) *A Review of Corporate Governance in UK Banks and Other*

Financial Industry Entities: Final Recommendations, London: The Walker Review Secretariat. Available at: www.hm-treasury.gov.uk/walker_review_information.htm (last accessed 7 May 2014).

Thomas, S. (1996) 'The privatization of the electricity supply industry', in J. Surrey (ed.) *The British Electricity Experiment. Privatization: The Record, the Issues, the Lessons.* London: Earthscan, pp. 40–66.

Waterson, M. (2015) 'Regulation and governance of public utilities', in Bailey, D., Cowling, K. and Tomlinson, P.R. *New Perspectives on Industrial Policy for a Modern Britain*, Oxford: Oxford University Press.

Vives, X. (2010) 'Competition and stability in banking', CEPR policy insight no. 50. Available at: www.cepr.org (last accessed 7 May 2014).

Williamson, O.E. (1975) *Markets and Hierarchies: Analysis and Antitrust Implications*, New York: The Free Press.

Williamson, O.E. (1985), *The Economic Institutions of Capitalism: Firms, Markets, Relational Contracting*, New York: The Free Press.

Wilson, J.R. (2008) 'Territorial competitiveness and development policy', Orkestra Working Paper Series in Territorial Competitiveness, no. 2008-02. Available at: www.orkestra.deusto.es (last accessed 7 May 2014).

Wren, C. (2001) 'The industrial policy of competitiveness: a review of recent developments in the UK', *Regional Studies*, 35: 847–860.

Yafeh, Y. (2000) 'Corporate governance in Japan: past performance and future prospects', *Oxford Review of Economic Policy*, 16 (2): 74–84.

Zeitlin, M. (1974) 'Corporate ownership and control: the large corporation and the capitalist class', *American Journal of Sociology*, 79: 1073–1193.

Index

3-k work 133

'Abenomics' 126
absorbtive capacity 13, 23; across firms 17, 18
Advanced Camera for Surveys 61
agency work 134, 135, 136
Air Force Cambridge Research Lab 61
Airvana 64
Alcatel 63
Alix Partners 184
Alvarez *et al.* 33
American International Group (AIG) 2
Ankur digital signal-processing (DSP) chip 106
Antonioli *et al.* 33
apparel industry 62, 86–7; declining employment 83
Archibugi *et al.* 33
Ascend Communications 63
ASEAN (Association of Southeast Asian Nations) 124, 147n11
Asian Financial Crisis (1997) 127
AT&T 62
auto industry: comparing North American and European conditions 2009-2014 179–83; industry conditions at onset of recession 170–2; mass production 58; outlook 183–5; policy response in Europe 176–9; policy response in North America 172–6
auto industry, Europe: capacity reduction 182; capacity utilization 182–3; comparison with North American auto industry conditions 2009-2014 179–83; cyclical pattern 171–2; declining sales 170, 171; economic recovery plan 177; European Commission communication 178–9; gap between capacity and demand 184; new vehicle registrations 178; outlook 183–5; policy response 176–9; production 181–2; sales and market share 179–80; scrappage programs 178; temporary framework for state aid 177–8
auto industry, India: banning of imports 100; boost to industry 102–3; capital equipment 112; cars as luxury products 100; commercial joint ventures 102–4, 106–7, 112; component manufacturing 100–1; components sector 108; flexible business model 112; fuel efficiency standards 107, 108; industrial governance 112; industrial strife 112; initial conditions 100–7; ISI (import substitution industrialization) 100–2, 106; Japanese industry practices 112; liberalization of 101; market forces 115–16; markets and industry structure 107–11; Maruti Udyog Limited (MUL) 102, 104, 107, 107–8, 112, 113, 114–15; middle-class demand 102, 108; people's car project 104; policy of unlimited production of commercial vehicles 100–1; pre-cut steel imports 108; regulation of 100; single model production 100; small car specialization 108; state support 114–15; subcontracting arrangements 112; support for passenger car segment 101
auto industry, North America: auto valley 172; capacity reduction 182; capacity utilization 182–3; comparison with European auto industry conditions 2009-2014 179–83; declining sales 170, 171; industrial policy 174–6; legacy costs 175; outlook 183; policy response to recession 172–6; production 181–2; reduced productive capacity 176; regional policy response to recession